Quick Guide to Molecular Diagnostics

uick Guide to Molecular Diagnostics

D. Hunter Best, PhD, FACMG
Assistant Professor of Pathology (Clinical)
University of Utah School of Medicine
Medical Director, Genetics Division
ARUP Laboratories
Salt Lake City, Utah

Elaine Lyon, PhD, FACMG
Associate Professor of Pathology (Clinical)
University of Utah School of Medicine
Medical Director, Genetics Division
ARUP Laboratories
Salt Lake City, Utah

Kristina A. Roberts, PhD
Department of Pathology
University of Utah
Genetics Division
ARUP Laboratories
Salt Lake City, Utah

Alexandra Valsamakis, MD, PhD
Associate Professor of Pathology
Department of Pathology
Division of Medical Microbiology
Johns Hopkins Hospital
Baltimore, Maryland

1850 K Street, NW, Suite 625
Washington, DC 20006

Quick Guide Series

Lorenz RG, Nahm MH—*Quick Guide to Autoimmune Disease Serology*
Chiasera JM, Hardy RW, Smith JA—*Quick Guide to Clinical Chemistry, Second Edition*
Marques MB, Fritsma GA—*Quick Guide to Coagulation Testing, Second Edition*
Winter WE, Bazydlo LA, Harris NS—*Quick Guide to Endocrinology*
Reddy V, Marques MB, Fritsma GA—*Quick Guide to Hematology Testing, Second Edition*
Datta P, Ejilemele AA, Peterson JR—*Quick Guide to Immunoassay Interference*
Fritsma GA, McGlasson DL—*Quick Guide to Laboratory Statistics and Quality Control*
Jones PM, Rakheja D—*Quick Guide to Organic Acid Interpretation*
Fritsma GA—*Quick Guide to Renal Disease Testing*
Marques MB, Fritsma MG—*Quick Guide to Transfusion Medicine*
Bennett A, Fritsma GA—*Quick Guide to Venipuncture, Revised Printing*

Second Printing 2017. AGS

Printed in the United States of America

Library of Congress Cataloging-in-Publication Data

Best, D. Hunter, author.
Quick guide to molecular diagnostics / D. Hunter Best, Elaine Lyon, Kristina A. Roberts, Alexandra Valsamakis.
p. ; cm. -- (Quick guide series)
Includes bibliographical references and index.
ISBN 978-1-59425-159-7 (alk. paper)
I. Lyon, Elaine, author. II. Roberts, Kristina A., author. III. Valsamakis, Alexandra, author. IV. American Association for Clinical Chemistry, publisher. V. Title. VI. Series: Quick guide series (American Association for Clinical Chemistry)
[DNLM: 1. Molecular Diagnostic Techniques--Handbooks. 2. Communicable Diseases--diagnosis--Handbooks. 3. Genetic Testing--Handbooks. 4. Neoplasms--diagnosis--Handbooks. QY 39]
RC582
616.07'9--dc23

2013022797

Contents

Figures and Tables

Figures

Tables

Techniques and Technologies

Molecular diagnostics is a constantly evolving field with many variations in technique and rapidly advancing technology. In this section, we briefly discuss techniques most commonly used in clinical molecular diagnostic laboratories. These can be divided into three categories:

1. **Extraction techniques:** Methods to isolate nucleic acids (DNA or RNA) from cells.
2. **Amplification techniques:** Methods to increase the copy number of a specific sequence of interest.
3. **Detection techniques:** Methods for detecting common mutations, sequence variants, and large deletions and duplications, as well as other types of genetic variation.

General methodologies, typical uses, advantages, and limitations are discussed for each technique.

Extraction Techniques

Obtaining high quality, purified DNA or RNA from a patient sample is often the first step in the molecular diagnostic process and can be critical to the success of subsequent assays. A variety of extraction methods are available and protocol selection should take into account both the sample source and the requirements of the assay to be performed. The needs of the laboratory (e.g., automated workflow) may also factor into method selection.

DNA Extraction

Many different techniques can be used to isolate genomic DNA (gDNA) from cells or tissues. In general, all of these methods involve the same three basic steps:

1. Tissue dissociation (if necessary) and cell lysis
2. Removal of proteins and other contaminants
3. Recovery of DNA

Cell lysis releases nucleic acids by disrupting cellular and nuclear membranes. Lysis must occur under conditions that will not damage the nucleic acid and is often accomplished by treating cells with an anionic detergent such as sodium dodecyl sulfate (SDS).

Protein removal and DNA recovery can be accomplished in many ways, each with its own advantages and limitations. Four common methods are described below.

- **Organic extraction:** Organic extraction is a traditional technique that uses organic solvents to extract contaminants from cell lysates. Following cell lysis, a mixture of phenol and chloroform is added to each sample. Hydrophobic contaminants collect in the organic phase, while DNA remains in the aqueous phase. Centrifugation settles the organic layer on the bottom, with the aqueous layer on top. Cellular debris collects as a white precipitate at the interface between the two layers. The upper aqueous phase can be collected and DNA recovered by alcohol precipitation.
    - **Advantage:** Generally yields relatively pure, high molecular weight DNA.
    - **Limitations:** (1) Phenol and chloroform are toxic so care must be taken to prevent exposure, and all waste should be disposed of in accordance with hazardous waste regulations. (2) Process is time-consuming, labor-intensive, and nearly impossible to automate, making it unsuitable for high-throughput applications. Because of these disadvantages, organic extraction methods are not routinely used in clinical laboratories.
- **Inorganic extraction:** Inorganic extraction, sometimes called "salting out," is a technique that uses high salt conditions to selectively precipitate proteins and other contaminants from cell lysates, while leaving the DNA in solution. After removal of these

precipitates by centrifugation, DNA can be recovered by alcohol precipitation.

- **Advantages:** (1) Reduces exposure to hazardous chemicals compared with organic extraction techniques. (2) DNA yields and purity are generally good.
- **Limitation:** Process can be time consuming and labor intensive.

- **Solid-phase extraction:** Solid-phase extraction techniques use solid, silica-based matrices to bind DNA. Many commercial kits use "spin columns" that contain a silica membrane and fit inside a microcentrifuge tube. When cell lysates are applied to the column, DNA adsorbs to the matrix. The immobilized DNA is washed and then eluted in a low salt buffer.
  - **Advantages:** (1) Extraction process is simple and fast. (2) Multiple samples can be simultaneously processed. (3) Standardized procedure leads to reproducible DNA quality.
  - **Limitations:** (1) More expensive per sample than other extraction methods. (2) DNA yield may be lower than with other extraction techniques.
- **Automated extraction:** Solid-phase extraction methods are employed by several robotic DNA isolation systems currently on the market. These instruments use magnetic silica beads to bind DNA. Magnets are then used to move and/or immobilize the beads as necessary, allowing the extraction process to be fully automated. Instruments are available with capacities ranging from eight to 96 samples to accommodate the throughput needs of different laboratories.
  - **Advantages:** (1) Reduces hands-on time required for DNA isolation. (2) Reduces chance of sample contamination or mix-up due to human error. (3) Automated procedure leads to reproducible DNA quality.
  - **Limitations:** (1) Instrument is expensive and must be properly maintained. (2) DNA yield may be lower than with other extraction techniques.

Common Sample Types for DNA Extraction

DNA can be extracted from a wide variety of biologic samples. Some of the most common diagnostic sample types are listed below, along with general collection guidelines. Specimen requirements may vary by assay and between diagnostic laboratories, so confirmation is recommended prior to ordering.

- **Whole blood:** Samples collected in ethylenediaminetetraacetic acid (EDTA) tubes are preferred, but acid citrate dextrose (ACD) tubes may also be acceptable. Heparin should be avoided as it binds to nucleic acids and is not removed by standard extraction techniques. Residual heparin can inhibit polymerase chain reaction (PCR) and other enzyme-based assays. DNA extracted from heparin samples should be diluted prior to PCR amplification.
- **Bone marrow:** Samples should be collected in EDTA tubes. Heparin should be avoided.
- **Buccal swab or saliva:** Buccal swabs and saliva kits provide a simple, noninvasive method of collecting cheek cells for DNA extraction. The quality and quantity of DNA from saliva samples are generally comparable with DNA extracted from whole blood. Buccal swabs generate lower DNA concentrations.
- **Amniotic fluid or chorionic villus sample:** Fetal genotype can be determined by testing DNA derived from amniocytes or chorionic villi. DNA can be extracted directly from amniotic fluid or chorionic villus sample (CVS), but this may not provide enough DNA for some assays. Alternatively, amniocytes or chorionic villi can be cultured prior to DNA extraction, resulting in higher DNA yields. It is recommended that backup cultures be maintained until molecular testing is complete.
- **Formalin-fixed paraffin-embedded tissue:** Formalin-fixed paraffin-embedded (FFPE) tumor tissue is frequently tested in molecular oncology assays. Thin sections are usually used for analysis and extraction, but sectioning is not required for very small samples. Tumor tissue may be microdissected away from

normal tissue using a separate stained serial section as a guide. FFPE tissue may also be used when no other tissue source is available for testing (e.g., when postmortem diagnostic testing is requested). DNA extracted from FFPE samples tends to be more fragmented than DNA from fresh tissues, and the amount of high molecular weight DNA may be significantly reduced in tissues fixed for more than 24 h. The significance of this fragmentation depends on the assay to be performed.

RNA Extraction

RNA is inherently less stable than DNA, and ribonucleases (RNases) are ubiquitous in the environment. Therefore, strict precautions must be taken to avoid RNA degradation. Ideally, fresh samples collected for RNA analysis should arrive in the testing laboratory within 48 h. RNase-free reagents and supplies should be used for all sample processing and, when possible, work should be carried out in a designated "RNase-free" area of the laboratory.

RNA can be isolated using organic and solid phase extraction methods similar to those described above for DNA. A couple of key differences are listed below:

- In organic extractions, cell lysis is often performed using a solution of phenol and guanidine isothiocyanate. Guanidine isothiocyanate aids in lysis, but it is also a strong denaturant of RNases. Immediately inactivating the intracellular RNases released during the lysis step is critical for successful RNA extraction.
- For many downstream applications, it is necessary to remove contaminating DNA from extracted RNA. DNase may be included in the extraction protocol (e.g., added during the lysis step) or directly added to the isolated RNA at the end of the procedure.

Common Sample Types for RNA Extraction

RNA can be extracted from a variety of biologic samples, but the RNA of interest must be expressed in the selected sample. Some common diagnostic sample types are listed below, along with general collection guidelines. Specimen requirements may vary by assay and between diagnostic laboratories, so confirmation is recommended prior to ordering.

- **Whole blood:** Samples should be collected in EDTA tubes. Heparin should be avoided as it binds to nucleic acids and is not removed by standard extraction techniques. Residual heparin can inhibit subsequent enzymatic reactions. RNA extracted from heparin samples should be diluted prior to use.
- **Bone marrow:** Samples should be collected in EDTA tubes. Heparin should be avoided.
- **FFPE tissue:** FFPE tumor tissue is frequently tested in molecular oncology assays. Thin sections are usually used for analysis and extraction, but sectioning is not required for very small samples. Tumor tissue may be microdissected away from normal tissue using a separate stained serial section as a guide. FFPE tissue may also be used when no other tissue source is available for testing (e.g., when postmortem diagnostic testing is requested).
- **Fresh frozen tissues:** A variety of fresh frozen tissues may be collected for analysis of tissue-specific expression patterns or detection of splice variants in genes with tissue-restricted expression.

Assessment of Nucleic Acid Quantity and Quality

It is often helpful to assess the quantity and/or quality of extracted nucleic acids before proceeding with molecular diagnostic testing. Spectrophotometry is a quick and easy way to evaluate the concentration and purity of extracted DNA and RNA.

- Nucleic acids show maximal absorption at a wavelength of 260 nm, while contaminating proteins absorb best at 280 nm. Measuring the optical density (OD) of an extracted sample at 260 nm and 280 nm allows for assessment of both nucleic acid concentration and purity.

- An OD_{260} of 1.0 corresponds to approximately 50 μg/mL of double-stranded DNA or 40 μg/mL of single-stranded RNA.
- The ratio of absorbance at 260 nm and 280 nm ($OD_{260/280}$) provides an estimate of nucleic acid purity. A pure DNA sample will have an $OD_{260/280}$ of approximately 1.8, while the $OD_{260/280}$ of pure RNA will be closer to 2.0. Lower ratios may indicate the presence of contaminating protein. Contamination with phenol or other organic solvents, as well as low pH, can also affect the $OD_{260/280}$ ratio.
- A number of "microvolume spectrophotometers" are now on the market. These instruments measure the absorbance of small sample volumes (1–2 μL), eliminating the need for time-consuming dilutions and minimizing wasted sample. Many of these devices include software that automatically calculates nucleic acid concentrations and $OD_{260/280}$ ratios.

Amplification Techniques

Amplification of extracted nucleic acids is often necessary to facilitate analysis of their nucleotide sequences or expression patterns. Here we describe some of the most common amplification techniques used in the molecular diagnostic laboratory, including both qualitative and quantitative methods. Although the focus of this section is on PCR-based amplification, it should be noted that a number of alternative amplification methods have been developed and are sometimes used in clinical settings.

Polymerase Chain Reaction

PCR is a simple yet robust molecular technique used to rapidly amplify DNA sequences *in vitro.* PCR uses DNA polymerase to copy a specific target sequence from a template DNA molecule. Successive synthesis cycles are performed, with each cycle generating new templates for the next round of amplification. This results in an exponential increase in the number of copies of the target sequence. The development of PCR in the 1980s revolutionized the field of molecular biology, and it is now an integral part of many molecular diagnostic tests.

PCR reactions require the following basic components:

- **Template DNA:** gDNA and complementary DNA derived from RNA (cDNA) are common templates in molecular diagnostics.
- **Thermostable DNA polymerase:** Thermostable polymerases maintain their activity throughout the repeated heating and cooling cycles of PCR. *Taq* polymerase, isolated from the thermophilic bacterium *Thermus aquaticus*, is the most commonly used PCR enzyme.
- **Oligonucleotide primers:** PCR primers are short, single-stranded DNA molecules complementary to the target of interest. Forward and reverse primers flank the region to be amplified.
- **Nucleotides:** The four deoxynucleotide triphosphates (dNTPs)—dATP, dTTP, dCTP, dGTP—are the building blocks of the DNA strands synthesized by PCR.
- **PCR reaction buffer:** PCR buffer provides optimal conditions for enzyme activity.

Each PCR cycle consists of three steps:

1. **Denaturation:** Template DNA is denatured by heating it to 94 °C or higher. This causes the strands of double-stranded DNA to separate from one another, producing the single-stranded DNA templates necessary for replication.
2. **Primer annealing:** Oligonucleotide primers hybridize to the denatured target DNA. Annealing temperature is specific to the PCR primers and reaction conditions. Primer sequences can be used to predict an appropriate annealing temperature, but optimization may be required.
3. **Primer extension:** Synthesis of the new DNA strand begins when the reaction is raised to the optimal temperature for the DNA polymerase (around 72 °C for most thermostable polymerases). New

DNA strands are synthesized from the 3' ends of the forward and reverse primers. Extension time depends on the polymerase used and the length of the DNA fragment being amplified.

These steps are automatically performed on a thermal cycler, which rapidly cycles through the required temperatures, holding each for the specified amount of time. At the end of the three-step cycle, the target region of each double-stranded DNA template has been replicated into two new double-stranded copies. These copies serve as the template DNA for the next PCR cycle, resulting in exponential amplification. PCR typically consists of 20–40 cycles with the exact number empirically determined.

- **Advantage:** PCR is a simple procedure with excellent sensitivity and specificity.
- **Limitations:** (1) Single nucleotide polymorphisms under the primer-binding site can lead to dropout of that allele. (2) Method is typically qualitative and, therefore, not used to analyze expression levels or copy number although PCR can be adapted to give quantitative information (see below). (3) Because of the sensitivity of PCR, extreme care must be taken to avoid cross-contamination between samples. In the clinical laboratory, separate preamplification and postamplification areas must be defined to prevent contamination of patient samples and/or reactions with previously amplified DNA.

Reverse Transcriptase-Polymerase Chain Reaction

Reverse transcription (RT) is the enzyme-mediated synthesis of a DNA molecule from an RNA template. The resulting DNA, known as complementary DNA or cDNA, can be used as a template for PCR amplification. RT followed by PCR is known as RT-PCR.

Although the details of RT protocols differ, the basic procedure involves incubating purified RNA with oligonucleotide primers, nucleotides (dNTPs), reverse transcriptase enzyme, RNase inhibitor, and RT buffer. Three different types of primers can be used in the RT reaction:

- **Oligo(dT):** Primers that selectively anneal to the poly(A) tails found on most messenger RNA (mRNA) molecules. Only polyadenylated RNAs will be reverse transcribed in an oligo(dT)-primed reaction.
- **Random hexamers:** A mixture of random hexanucleotide primers that anneal to sequences throughout the target RNA, allowing for reverse transcription of both polyadenylated and nonpolyadenylated RNAs.
- **Sequence-specific:** Primers that hybridize to a specified gene sequence and result in reverse transcription of a specific mRNA. When using sequence-specific primers, a new RT reaction must be performed for each gene of interest.

The selected primers anneal to the RNA, and reverse transcriptase synthesizes cDNA from the end of the primer, using the RNA sequence as a template. Upon completion of the RT reaction, samples are heated to inactivate the reverse transcriptase. Template RNA can then be removed by incubating with RNase H, which specifically degrades the RNA in RNA:DNA hybrids. This step may increase the sensitivity of subsequent PCR reactions, especially for long templates. If oligo(dT) or random hexamer primers are used, the resulting cDNA can be used in multiple PCR reactions, allowing for analysis of more than one gene.

"One Step" RT-PCR kits are available commercially that combine the RT and PCR steps into a single reaction. Although "One Step" kits simplify the RT-PCR process and may help prevent cross-contamination between samples, they can only be carried out with sequence-specific primers. It is not possible to use oligo(dT) or random hexamer primers in a one-step RT-PCR reaction.

- **Advantages:** (1) RT-PCR can be used to detect and analyze relatively low-abundance transcripts. (2) Procedure is less complex than traditional Northern blotting techniques and can be used to study samples with limited starting material. (3) Can be used to analyze mRNA splice variants and detect gene fusion products (e.g., BCR/*abl*).

- **Limitations:** (1) RT followed by conventional PCR cannot be used to accurately quantitate RNA transcripts. (2) RNA is inherently unstable so care must be taken to prevent degradation. The quality and purity of the RNA template is crucial to the success of RT-PCR.

Quantitative PCR and Quantitative RT-PCR

Standard, qualitative PCR methods indicate whether a particular target sequence is present in the template DNA and permit analysis of the amplified products at the nucleotide level. However, in some situations, it is important to quantify the amount of target sequence in a particular sample. Quantitative PCR (qPCR), also called real-time PCR, combines PCR amplification and detection into a single step, allowing for accurate quantification of target copy number. Fluorescent dyes are used to label PCR products so that their accumulation can be monitored in real time.

Analysis of standard PCR products occurs only at the end point of amplification when product accumulation has already plateaued due to exhaustion of reaction components and competition for PCR primers during the annealing step. When examined at this end point, the amount of PCR product does not reliably correlate with the quantity of starting template. By contrast, real-time PCR allows reactions to be monitored throughout the exponential phase of PCR, when the amount of PCR amplicon is directly proportional to the amount of starting material. Fluorescence is measured after each round of amplification and is directly proportional to the amount of amplicon present.

Fluorescent labeling of quantitative PCR products can be accomplished in a number of different ways. Two general strategies are discussed below.

- **Fluorescent DNA-binding dyes**
  - Fluorescent DNA-binding dyes (e.g., SYBR Green) provide a nonspecific, but relatively simple, method of labeling qPCR products. These dyes show little fluorescence when free in solution, but their fluorescence increases dramatically when

bound to double-stranded DNA. Because the dyes bind to all double-stranded DNA, including nonspecific PCR products and primer-dimers, PCR conditions must be optimized to ensure that the reaction generates only the intended product.

- Amplification specificity can be verified at the end of each run by melting curve analysis. Melting curves are generated by slowly increasing the reaction temperature to denaturing levels while continuously monitoring fluorescence. Since product length and sequence both affect melting temperature, the resulting melting curve can be used to assess amplicon homogeneity.
- DNA-binding dyes cannot be used for multiplex assays since only one type of amplicon can be detected per reaction.

- **Target-specific hybridization probes**
  - Fluorescently labeled hybridization probes (e.g., TaqMan, Molecular Beacons) allow for specific detection of a particular qPCR amplicon. These single-stranded oligonucleotides are labeled with a reporter molecule that fluoresces only after hybridization of the probe with its DNA target. This is often achieved by distancing the fluorescent reporter from a quencher molecule that is also attached to the probe. The exact mechanism by which this separation occurs depends on the probe technology being used.
  - Fluorescence resonance energy transfer (FRET) probes can also be used for qPCR. FRET probes consist of two oligonucleotides designed to hybridize to adjacent regions on the target DNA, each labeled with a different fluorophore. These two fluorophores interact only when both probes are bound to their targets, bringing the dyes into close proximity. Energy transfer from the donor to the acceptor causes the acceptor to fluoresce at a specific, detectable wavelength.
  - Hybridization probes reduce the need for stringent PCR optimization because detection is limited to the target of interest. Multiplexing is possible because, unlike DNA-binding dyes, probe-based detection methods can distinguish between different targets in the same reaction.

Two different methods can be used to quantify qPCR products:

- **Relative quantification** compares the gene of interest to an internal reference gene. This method can be used to determine fold-differences in expression of the target gene.
- **Absolute quantification** is achieved by comparing the fluorescence of the unknown test sample to the fluorescence of the same target in serially diluted DNA standards. This method calculates the number of target DNA molecules present in the initial sample.

- **Advantages:** (1) qPCR has a wide dynamic range so samples with high and low initial copy numbers can be accurately quantified in the same run. (2) Sample analysis is fast since no post-PCR processing steps are required. (3) Analysis occurs within a closed-tube system, reducing the risk of sample mix-up, sample contamination, and contamination of the laboratory environment with PCR amplicon.
- **Limitations:** (1) Equipment and fluorescent dyes can be relatively expensive. (2) Data quality and accuracy are highly dependent on the quality of the template DNA. (3) For relative quantification, 1:2 and 2:3 copy number ratios require careful optimization.

Detection Techniques

Molecular diagnostic laboratories use a wide variety of techniques and technologies to detect disease-causing mutations. Selection of an appropriate detection technique must take into account the types of mutations expected and the purpose of the assay. In this section, we discuss techniques used for assaying specific mutations, generating sequence data, and detecting large deletions and duplications.

- **Targeted mutation analysis:** Detection of specific variants of known clinical importance. Variants not targeted by the assay will not be detected.

- **Sequence analysis:** Detection of all variants within a specified region, including pathogenic mutations, benign polymorphisms, and variants of uncertain clinical significance.
- **Deletion/duplication analysis:** Detection of large (whole exon or larger) deletions or duplications not detectable by sequencing.

Separation of Nucleic Acids by Size

Many molecular diagnostic techniques require separation of nucleic acids by size. Electrophoresis uses an electric current to move charged molecules through a porous matrix, which impedes their migration and results in size-based separation. The migration speed of a nucleic acid fragment is inversely related to its length, as short fragments pass through the pores of the matrix more easily than longer ones. In the molecular diagnostic laboratory, electrophoresis is most frequently performed in slab gels or polymer-filled capillaries.

- **Agarose gel electrophoresis** is a relatively simple method of separating nucleic acids by size.
  - Nucleic acid migration takes place in a slab gel composed of agarose, a polysaccharide polymer derived from seaweed. The concentration of agarose determines the size of the pores in the gel; therefore, it is chosen based on the approximate size of the nucleic acid fragments to be resolved. Small fragments are resolved in more concentrated agarose gels (e.g., 2%–3%), while larger fragments are better resolved in gels with lower agarose concentrations (e.g., 0.5%–1%).
  - Samples are loaded in wells molded into the gel so that each migrates in its own lane. Gels are stained with chemicals, such as ethidium bromide, which bind to nucleic acids and fluoresce when exposed to ultraviolet light. Separated nucleic acids appear as bands on the gel and can be compared to known molecular weight markers (run in an adjacent lane) to determine their approximate sizes.

- **Advantages:** (1) Easy to perform. (2) Required equipment is fairly inexpensive.
- **Limitations:** (1) Process is hard to automate and throughput is limited. (2) Size determination is approximate and may be affected by lane-to-lane variation across the gel. (3) Longer fragments do not migrate as far as shorter fragments, resulting in lower resolution of high molecular weight nucleic acids. (4) Very small nucleic acids cannot be resolved on agarose gels. Polyacrylamide gels can be used to separate small fragments and have very high resolution, but clinical laboratories are more likely to use capillary electrophoresis for this type of separation.

- **Capillary electrophoresis** is a powerful technology that allows for automated separation of nucleic acids by size. Capillary electrophoresis (CE) is an integral part of many molecular diagnostic techniques, including targeted mutation analysis, Sanger sequencing, multiplex ligation-dependent probe amplification (MLPA) detection of deletions and duplications, and short tandem repeat (STR)-based identity testing.
  - Separation is performed in a thin, polymer-filled capillary column made of fused silica. Fluorescently labeled nucleic acids are injected into the capillary by applying a transient high voltage charge that draws the fragments into the end of the column. Under the force of an electric current, fragments then migrate through the capillary, with small molecules moving through the polymer faster than larger molecules.
  - Shortly before reaching the end of the capillary, separated nucleic acids pass by a detector that uses a laser beam to excite the fluorescent dye attached to each fragment. The light emitted by these dyes is captured by an optical detection device and correlated with migration time.
  - Because fluorescent dyes emit light at different wavelengths, CE detectors can monitor multiple dyes in a single capillary injection. This enables simultaneous analysis of multiple independent markers and allows a labeled size standard to be included in each run.

- Selection of an appropriate CE system is based on the throughput needs of the laboratory and the specific type of separation to be performed. CE analyzers vary by (a) capillary length, (b) polymer type, (c) number of capillaries, and (d) sample capacity (i.e., the number of samples that can be queued for automated injection). The combination of capillary length and polymer type determine the resolution and also influence sample run time. Run time, capillary number, and sample capacity all affect throughput.

- **Advantages:** (1) Injection, separation, and detection steps can all be automated. (2) Provides excellent sensitivity and resolution. (3) Time at which fragments elute can be precisely determined and compared to elution times of internal size standards. (4) Fragments are detected only after experiencing the separation power of the entire column so resolution of longer fragments is comparable to that of shorter ones.
- **Limitations:** (1) CE instruments are expensive and capillaries must be replaced at regular intervals. (2) Fluorescent labeling of samples can be costly. (3) Samples must be completely denatured before injection, as any secondary structure will affect migration speed.

Detection of Targeted Mutations

Targeted mutation analysis is used to quickly detect specific variants of known clinical significance. Targeted techniques are useful for analyzing:

- Common mutations responsible for a significant proportion of a given disease
- Mutations that occur at high frequency in specific ethnic populations (e.g., for carrier screening)
- Recurrent mutations found in cancer cells that may influence prognosis, treatment options, etc.

In some cases, targeted testing may be used as a first-tier, low-cost test to detect the most common disease-causing mutations. If targeted testing is negative or shows only one mutation when two are expected (e.g., in the case of recessive diseases), sequencing or deletion/duplication tests may be added to look for additional causative mutations.

There are many methods available for detection of targeted mutations. Although each has its own advantages and limitations, method selection is often influenced by the techniques and instrumentation already being used in a particular clinical laboratory. The general advantages and limitations of targeted mutation testing, as well as a few common technologies, are described below.

- **Advantages:** (1) Procedures are generally designed to be quick and relatively inexpensive. (2) Many platforms permit multiplexing so that multiple mutations can be assayed at once. Level of multiplexing and throughput depend on the specific technology employed. (3) Only targeted mutations are assayed to avoid detection of mild or uncertain variants.
- **Limitations:** (1) Nontargeted variants located under the primers or probes may interfere with the assay. Depending on assay design, this may result in false positives, false negatives, or inconclusive results that require additional follow-up by the laboratory. (2) Only targeted mutations will be detected so additional testing may be required to detect rare pathogenic mutations. (3) It is often necessary to design separate reactions or probes for each targeted mutation. (4) Careful design and optimization are necessary for mutations in close proximity to one other and for mutations in close proximity to common polymorphisms.

- **Melting curve analysis**
  Melting curve analysis exploits differences in melting temperature between perfectly matched and mismatched strands of DNA. In this method, fluorescently labeled probes are designed to be a perfect match for the wild-type sequence of interest. Probes are hybridized to the template DNA and then DNA melting (i.e., slowly raising

the reaction temperature to denaturing levels) is performed in the presence of fluorescence monitoring. When a mutation is present, the mismatch between the probe and template DNA lowers the melting temperature of the double-stranded DNA:probe complex. Thus, wildtype and mutant alleles generate different melting curves. Individuals heterozygous for a mutation will show a melting curve that corresponds to both the wildtype and mutant alleles.

- **Oligonucleotide ligation assay**
  Oligonucleotide ligation assay (OLA) utilizes two oligonucleotide probes designed to hybridize to adjacent sites in a target DNA, such that there is no gap between them. The nucleotide to be assayed is adjacent to the join site of the probes. If the ends of the probes are perfectly matched with the target sequence, DNA ligase will covalently bond them together, creating a single molecule. If the probe ends are not perfectly matched, ligation will not occur. When PCR is then performed with one primer that hybridizes to each oligonucleotide, only ligated oligonucleotides will be amplified. Using fluorescently labeled primers, PCR products can be detected by capillary electrophoresis.
- **Single-base extension**
  Single-base extension (SBE), also known as single nucleotide extension (SNE), can detect approximately 20 mutations in a single reaction. A single long-range PCR or a multiplexed PCR amplifies the region(s) of interest and is followed by analysis with a multiplexed set of extension primers that anneal immediately adjacent to the mutations of interest. A second, linear amplification incorporates the next nucleotide at the mutation position. The incorporated base is a fluorescently labeled dideoxynucleoside (ddNTP), which prevents further extension. Products are then separated by capillary electrophoresis to determine which nucleotide was incorporated at each location.
- **Allele-specific oligonucleotide hybridization**
  Under stringent hybridization conditions, a short oligonucleotide (e.g., 15–20 nucleotides) will only hybridize to a perfectly matched

sequence. In an allele-specific oligonucleotide (ASO) assay, two oligonucleotides are synthesized for each targeted mutation—one is complementary to the normal sequence and the other to the mutated sequence. These oligonucleotides are incubated with target DNA and probe:DNA binding is detected. ASO techniques can be used on a number of different platforms, including oligonucleotide microarrays, microplate assays, and dot blots.

- **MALDI-TOF mass spectrometry**
  A variety of PCR methods, such as SBE, can be adopted for mass spectrometry (MS) applications, which can detect up to a few hundred mutations. Similar to other methods, an initial PCR amplifies the targeted region, followed by a primer extension step. After removing unincorporated nucleotides and salts that may interfere with analysis, samples are ionized, separated based on their mass-to-charge ratios, and detected after laser desorption with a nitrogen laser. The mass differences resulting from nucleotide base changes can be detected. MS can detect low levels of sequence variants, making it useful for analyzing mosaic samples, heteroplasmy in mitochondrial DNA, and somatic changes in tumor samples mixed with normal cells. The throughput of this method is limited by the level of multiplexing achievable in the SBE reaction so several reactions may be required to maximize the MS capabilities.

Detection of Sequence Variants

Many molecular diagnostic tests require analysis of the complete DNA sequence of a particular region of interest (e.g., the coding exons of a disease gene). Sequence information is necessary when mutations are spread across a gene or region and when many different mutations are responsible for disease. Sequencing is also able to discover novel mutations that have not previously been described. Depending on the disease and specific clinical situation, sequencing may be a first-tier test or may be used as a follow-up to targeted mutation analysis.

- **Sanger sequencing**, also known as dideoxy chain termination sequencing, is a modification of the DNA replication process and relies on random inhibition of chain elongation to generate DNA strands of various lengths that can be separated by size. The dye-terminator method of Sanger sequencing is now the mainstay of automated DNA sequencing.

  Sanger sequencing reactions include:
  - Single-stranded DNA template
  - Oligonucleotide sequencing primer
  - DNA polymerase
  - dNTPs
  - Fluorescently labeled ddNTPs

  ddNTPs lack the hydroxyl group found on the 3' carbon of dNTPs; therefore, they cannot form phosphodiester bonds at their 3' ends. Once a ddNTP is added to a growing strand of DNA, no further nucleotides can be added.

  Because both dNTPs and ddNTPs are included in the reaction mixture, an individual DNA strand will grow until a ddNTP is randomly incorporated, resulting in chain termination. The ratio of dNTPs:ddNTPs is critical for quality Sanger sequencing. If the concentration of ddNTPs is too high, termination will too often occur early in the template. If the ddNTP concentration is too low, termination will occur infrequently or not at all. Commercial dNTP:ddNTP mixes have been carefully optimized to ensure termination occurs throughout the length of the DNA template.

  In dye-terminator sequencing, each ddNTP is labeled with a different fluorescent dye. Therefore, fragments are color-coded based on their incorporated terminal ddNTP and can be sorted by size on a CE instrument. The peaks of fluorescence detected by CE can be read in order to determine the nucleotide sequence of the template DNA. If an individual is heterozygous at a particular nucleotide, two peaks of different colors will be seen directly on top of one another at that location.

- **Advantages:** (1) Generates sequence data with single nucleotide resolution. (2) Detects all sequence variants (i.e., single nucleotide changes and small insertions/deletions) in the targeted region. (3) Read lengths are often long enough to cover an entire exon.
- **Limitations:** (1) Cannot detect large (e.g., exonic) deletions and duplications. (2) Rare variants under the primer may cause allele dropout. For this reason, bidirectional sequencing (i.e., sequencing with both a forward and reverse primer) is strongly recommended and is standard practice in clinical diagnostic laboratories. (3) Cannot reliably detect low levels of mosaicism. (4) May detect variants that have not been previously identified and whose clinical significance is uncertain.

- **Pyrosequencing** is a "sequencing-by-synthesis" technique that records the sequence of a DNA strand while it is being synthesized. This method uses a cocktail of enzymes to monitor nucleotide incorporation, exploiting the natural release of pyrophosphate (PPi) that occurs each time a dNTP is successfully added to a growing DNA chain. Pyrosequencing was initially developed to assay single nucleotide polymorphisms, but sequencing of longer fragments can be accomplished by performing a series of successive pyrosequencing reactions. This is known as iterative pyrosequencing.

  Pyrosequencing reactions include:
  - Single-stranded DNA template
  - Oligonucleotide sequencing primer
  - DNA polymerase
  - ATP sulfurylase
  - Adenosine 5′ phosphosulfate (APS)
  - Luciferase
  - Luciferin
  - Apyrase

Once the reaction mixture is assembled and the sequencing primer has annealed to the template DNA, dNTPs are added one at a time.

1. If the dNTP added to the reaction is complementary to the next base in the template strand, DNA polymerase catalyzes its incorporation and PPi is released.
2. ATP sulfurylase quantitatively converts PPi to ATP in the presence of APS.
3. ATP drives the luciferase-mediated conversion of luciferin to oxyluciferin. Oxyluciferin generates visible light in amounts that are proportional to the amount of ATP.
4. Light is detected by a charge-coupled device chip and displayed as a peak in the raw data output, known as a Pyrogram. Peak height is proportional to the number of nucleotides incorporated.
5. Apyrase continuously degrades unincorporated nucleotides and ATP. If the dNTP added to the reaction mixture was not needed for the next synthesis step, no light is produced and apyrase will degrade the dNTP before the next nucleotide is added.

Upon completion of this iterative process, nucleotide sequence can be determined by analyzing the order and heights of peaks on the Pyrogram. A repeated nucleotide (e.g., GG) will appear as a peak that is twice the height of a single nucleotide peak.

- **Advantages:** (1) Sequence data are quantitative and are, therefore, ideal for measuring relative amounts of each allele. This is especially useful for detecting mixed genotypes in heterogeneous samples (e.g., tumor cells mixed with normal cells). (2) Sensitivity enables detection and quantification of mutations in even minor subpopulations. (3) Method is fast, relatively inexpensive, and easily automated.
- **Limitations:** (1) Works best for short-to-medium length sequences. (2) Sequence downstream of an inserted or deleted

nucleotide may be difficult to analyze. (3) Special techniques or software may be required to determine the exact number of nucleotides incorporated in long, homopolymeric regions (e.g., >6 identical nucleotides).

- **Next-generation sequencing (NGS)** is a relatively new sequencing technology that has already revolutionized genetic research and is now making its way into the clinical diagnostic laboratory.
    - Compared with Sanger sequencing, NGS allows for much higher throughput at significantly lower cost per nucleotide sequenced. While Sanger sequencing analyzes one target (e.g., exon) per reaction, NGS can sequence hundreds to thousands of targets in a single reaction system. NGS is often referred to as "massively parallel" sequencing because millions of sequencing reactions are simultaneously monitored, generating a tremendous amount of sequence data.
    - NGS can be performed on a variety of platforms, each using a slightly different sequencing technology. The details of these systems are beyond the scope of this chapter.
    - All NGS platforms yield much shorter sequence reads than Sanger sequencing and significant computational analysis is required to align these short reads to a reference genome. Once these sequences are aligned, the number of reads at each nucleotide (known as "read depth" or "depth of coverage") and a variety of other quality metrics are calculated.
    - After alignment, variants are bioinformatically annotated and filtered to help the laboratory focus on those most likely to be causative in a particular patient. Filters may include population frequency, suspected inheritance pattern, variants known to be benign or pathogenic, genes relevant to the patient phenotype, etc. Even after filtering is complete, substantial hands-on time is required to investigate and determine the clinical significance of each variant.

NGS is currently used for three main types of tests:

- **Multigene panels:** NGS gene panels simultaneously sequence all exons in a targeted set of genes. Genes that cause the same disease (e.g., Noonan syndrome) or result in similar/overlapping phenotypes (e.g., mitochondrial disorders, cardiomyopathies) are often grouped together on a panel. Gene panels are generally more cost effective than sequencing multiple genes individually and can simplify the molecular diagnostic process for clinicians. NGS panels may also target genes likely to be mutated in cancer cells for use as a diagnostic or prognostic indicator.
- **Exome sequencing:** Exome analysis involves sequencing the exons of all known protein-coding genes. Introns and intergenic regions are not sequenced. Exome sequencing is currently being used for diagnosis in patients with complex phenotypes that do not fit any known disorder and/or in whom previous testing has failed to yield a molecular diagnosis. Additional family members (e.g., parents, affected siblings, unaffected siblings) may be sequenced as controls to aid in variant filtering.
- **Whole genome sequencing:** Whole genome analysis involves sequencing all gDNA, including introns and intergenic regions. While it is possible to perform whole genome sequencing (WGS) in the clinical laboratory, it is more often utilized in research settings (e.g., for gene discovery projects). WGS detects many more variants than exome sequencing, but deep intronic and intergenic variants can be extremely difficult to classify because the function and/or significance of these regions is often unknown.

- **Advantages:** (1) Less expensive per nucleotide than Sanger sequencing. (2) Generates sequencing data for many targets at once. (3) Can be used in clinical cases with no obvious candidate gene(s).
- **Limitations:** (1) Requires significant computational and manual analysis. (2) Likely to detect many variants of uncertain clinical

significance. (3) Cannot currently detect large deletions and duplications. However, new algorithms to facilitate this type of analysis are in development. (4) Mutations should be confirmed by a secondary method (e.g., Sanger sequencing) before reporting.

Detection of Large Deletions and Duplications

The sequencing techniques described above detect nucleotide substitutions and small insertions/deletions, but generally cannot detect larger deletions and duplications. For many disorders, deletion/duplication analysis significantly increases the mutation detection rate over sequencing alone. Diseases caused by common or recurrent deletions with known breakpoints (e.g., α-thalassemia) can be diagnosed by using PCR to amplify over the breakpoint and then detecting the presence or absence of a product. However, specialized technologies must be used to identify deletions and duplications with unknown breakpoints.

- **MLPA** is a multiplex method for detecting abnormal copy number of gDNA targets. This technique is often used to systematically look for deletions or duplications in multi-exon genes and can assay up to 50 different targets per reaction. MLPA utilizes a multiplexed OLA that allows for simultaneous PCR amplification of all bound probes.

  For each targeted region (e.g., an exon or nearby intronic sequence), two oligonucleotides are designed to hybridize to adjacent sites on the DNA. By contrast to traditional OLA, the join site is positioned at an invariant nucleotide so that ligation will always occur if the template sequence is present. Each probe also includes a primer-binding site, as well as a "stuffer sequence" that determines the length of the amplicon produced during the PCR step. Identical primer-binding sequences are used for all oligonucleotide pairs in the reaction.

  MLPA consists of the following steps:

  1. **Denaturation:** Template DNA is heated, causing the strands of double-stranded DNA to separate from one another.

2. **Probe annealing:** A mixture of oligonucleotide probes (e.g., targeting all exons of a given gene or genes) is hybridized to the denatured DNA.
3. **Ligation:** DNA ligase covalently joins adjacent probes, sealing the gap between them. Successful ligation creates a single, PCR-amplifiable molecule.
4. **PCR:** PCR is carried out using primers complementary to binding sites on the MLPA probes. Because each probe pair includes the same primer-binding sequences, only one set of primers is necessary for amplification. Ligated probes, not the template DNA sequences, are amplified during the PCR.
5. **CE:** PCR products are separated by size on a CE instrument and detected by the fluorescent label attached to one primer. The amount of fluorescence generated by a particular amplicon is proportional to the amount of starting target DNA.
6. **Data analysis and interpretation:** CE peak intensities can be normalized using a variety of algorithms or software packages and compared to control probes located outside the gene or region on interest. Appropriate control probes are included in each MLPA reaction.

- **Advantages:** (1) Can detect deletions or duplications of all exons in a gene simultaneously. Very large genes may require multiple reactions, but small genes can be grouped together into a single reaction. (2) Requires only a thermal cycler and CE instrument. (3) A wide range of clinically relevant MLPA kits are commercially available.
- **Limitations:** (1) Assay is sensitive to DNA quality and quantity so it is best to run samples in duplicate and be suspicious of discordant results. (2) Rare mutations under an MLPA probe may affect probe binding and appear as a deletion of the entire exon. Therefore, all apparent single exon deletions should be investigated further before reporting. Sequencing of the probe binding site and qPCR can help determine whether the exon

is truly deleted. (3) MLPA probes are complex and each one needs to be carefully designed so clinical laboratories generally buy commercial MLPA kits. Kits are currently only manufactured by a single company.

- **Hybridization-based microarray** technologies can be used to detect copy number changes in gDNA samples. These technologies include comparative genomic hybridization (CGH) microarrays and single nucleotide polymorphism (SNP) microarrays, and are often referred to as genomic microarrays, cytogenomic microarrays, or chromosomal microarrays.

  Microarrays detect genomic gains or losses by hybridizing DNA samples onto a set of probes representing specific regions of the genome. These probes are attached to a solid surface, known as an array or chip. A single array can contain hundreds of thousands to millions of probes. Microarrays can be designed to interrogate the entire genome or to focus on a specific region of interest.

  - **Genome-wide arrays:** Genome-wide arrays contain probes across all chromosomes, but they often have higher probe density in regions with annotated genes. These arrays can be used to screen for deletions or duplications anywhere in the genome that may be responsible for a patient's phenotype. Genome-wide arrays can also be used for prenatal diagnosis and for analyzing products of conception. In addition, these arrays are now being utilized to catalog changes in cancer cells that may have diagnostic or prognostic value.
  - **Targeted, high-resolution arrays:** Specialized arrays may target a particular chromosome or a specific set of genes at higher resolution (i.e., with more densely packed probes) than genome-wide arrays. Targeted arrays may be used in conjunction with NGS gene panels to provide comprehensive analysis of selected genes.

  CGH platforms use comparative hybridization of a test sample and a well-characterized reference sample to detect gains or losses in the test sample. Patient and reference DNAs are labeled with different fluorescent dyes, denatured, and then hybridized to the array. Fluorescence is monitored and the signal ratio for each probe (i.e., the ratio of patient-to-reference fluorescence) provides information about copy number variation (CNV) in the patient.

SNP platforms detect single nucleotide polymorphisms throughout the genome using probes designed to both the wildtype and variant alleles. SNP arrays often also contain copy number probes to provide even coverage across the genome and improve CNV detection. On these platforms, only the patient DNA sample is hybridized to the microarray and the resulting signal intensities are compared to those of a collection of previously analyzed reference samples. Arrays that include both CNV and SNP probes result in two complementary readouts, which can help facilitate data analysis. Unlike CGH arrays, SNP microarrays can detect copy-neutral loss of heterozygosity, as well as copy number variation.

- **Advantages:** (1) Targeted arrays can analyze many more targets at one time and have higher resolution than MLPA. (2) Genome-wide arrays allow for quick detection of CNVs in patients with a constellation of findings. It is not necessary to identify a gene or region of interest before performing the test. (3) Microarrays can detect CNVs that are below the resolution of fluorescent *in situ* hybridization (FISH) technologies. (4) Arrays that include SNP probes can detect long stretches of homozygosity that may be indicative of uniparental disomy or consanguinity, both of which increase the risk for autosomal recessive conditions.
- **Limitations:** (1) Initial microarray production can be expensive so design updates are limited. (2) Requires more specialized instrumentation than MLPA. (3) In most cases, CGH microarray is incapable of detecting small sequence alterations. (4) Microarrays will not identify balanced chromosomal rearrangements in the patient. (5) Arrays may detect CNVs of uncertain clinical significance.

- **FISH** is a cytogenetic technique that uses fluorescently labeled probes to detect the presence or absence of targeted DNA sequences on chromosomes. FISH can also be used to localize a sequence to a specific chromosome or to determine the relative locations of two sequences. The signals from hybridized FISH probes are analyzed by fluorescence microscopy. FISH is often utilized for:
  - **Detection of large deletions or duplications:** Counting the number of fluorescent signals per cell indicates whether the expected copy number is present.
  - **Detection and analysis of chromosomal translocations:** Probes targeted to breakpoint regions are either brought together or split apart in the presence of a specific translocation.

FISH probes targeted to a unique DNA sequence are often generated from bacterial artificial chromosomes (BACs) and range in size from approximately 1 Kb to more than 1 Mb. Unique sequence probes can also be generated by long-range PCR if a suitable BAC clone is not available. Probes targeted to repetitive sequences, such as those found at chromosome centromeres, can be much smaller than unique sequence probes. There is a wide array of commercially available FISH probes, but probes should be validated by the clinical laboratory prior to their use.

There are two types of FISH, which differ based on the phase of the cell cycle the cells are in at the time of the test. The procedure is similar in both cases and selection is based on the purpose of the test, the sample type available, and the required turnaround time.

- **Metaphase FISH:** Metaphase FISH examines cells that are currently dividing and in the metaphase stage of the cell cycle. During metaphase, chromosomes are condensed and can be individually distinguished. Therefore, sequences detected by metaphase FISH can be localized to a specific chromosome. Because cells must be actively dividing, metaphase FISH can only be performed on cells that will grow in culture.
- **Interphase FISH:** Cells examined by interphase FISH do not need to be actively dividing and are in the interphase stage of the

cell cycle. During interphase, chromosomes are not condensed and, therefore, cannot be distinguished from one another. Because cells do not need to be dividing, interphase FISH can be performed on a wider variety of sample types, including frozen or FFPE tissues. Interphase FISH tends to be faster than metaphase FISH because cell culture is not required.

- **Advantages:** (1) Less expensive than other methods for detecting deletions, duplications, and translocations. (2) Especially for interphase FISH, turnaround time is fast.
- **Limitations:** (1) Due to the size of the probes, cannot be used to detect exon-level deletions and duplications. (2) FISH probes must be targeted to the specific deletion, duplication, or translocation suspected in a patient. FISH cannot be used as a screen for deletions/duplications. (3) Some cell types are very difficult to grow in culture. Culturing cells may lead to selective amplification of certain clonal populations.

Suggested Reading

Alkan C, Coe BP, Eichler EE. Genome structural variation discovery and genotyping. Nat Rev Genet 2011;12:363–76.

Altmann A, Weber P, Bader D, Preuss M, Binder EB, Müller-Myhsok B. A beginners guide to SNP calling from high-throughput DNA-sequencing data. Hum Genet 2012:1541–54.

Bamshad MJ, Ng SB, Bigham AW, et al. Exome sequencing as a tool for Mendelian disease gene discovery. Nat Rev Genet 2011;12:745–55.

Bonin S, Stanta G. Nucleic acid extraction methods from fixed and paraffin-embedded tissues in cancer diagnostics. Expert Rev Mol Diagn 2013;13:271–82.

Brady PD, Vermeesch JR. Genomic microarrays: a technology overview. Prenat Diagn 2012;32:336–43.

Chen X, Sullivan PF. Single nucleotide polymorphism genotyping: biochemistry, protocol, cost and throughput. Pharmacogenomics J 2003;3:77–96.

Fakhrai-Rad H, Pourmand N, Ronaghi M. Pyrosequencing: an accurate detection platform for single nucleotide polymorphisms. Hum Mutat 2002;19:479–85.

Kozlowski P, Jasinska AJ, Kwiatkowski DJ. New applications and developments in the use of multiplex ligation-dependent probe amplification. Electrophoresis 2008;29:4627–36.

Miller DT, Adam MP, Aradhya S, et al. Consensus statement: chromosomal microarray is a first-tier clinical diagnostic test for individuals with developmental disabilities or congenital anomalies. Am J Hum Genet 2010;86:749–64.

Mullis K, Faloona F, Scharf S, Saiki R, Horn G, Erlich H. Specific enzymatic amplification of DNA in vitro: the polymerase chain reaction. Cold Spring Harbor Symp Quant Biol 1986;51:263–73.

Ronaghi M. Pyrosequencing sheds light on DNA sequencing. Genome Res 2001;11:3–11.

Ronaghi M, Karamohamed S, Pettersson B, Uhlén M, Nyrén P. Real-time DNA sequencing using detection of pyrophosphate release. Anal Biochem 1996;242:84–9.

Sanger F, Nicklen S, Coulson AR. DNA sequencing with chain-terminating inhibitors. Proc Natl Acad Sci U S A 1977;74:5463–7.

Schaaf CP, Wiszniewska J, Beaudet AL. Copy number and SNP arrays in clinical diagnostics. Annu Rev Genomics Hum Genet 2011;12:25–51.

Schouten JP, McElgunn CJ, Waaijer R, Zwijnenburg D, Diepvens F, Pals G. Relative quantification of 40 nucleic acid sequences by multiplex ligation-dependent probe amplification. Nucleic Acids Res 2002;30:e57.

Shendure J, Ji H. Next-generation DNA sequencing. Nat Biotechnol 2008;26:1135–45.

Smith LM, Sanders JZ, Kaiser RJ, et al. Fluorescence detection in automated DNA sequence analysis. Nature 1986;321:674–9.

Tiu RV, Gondek LP, O'Keefe CL, et al. Prognostic impact of SNP array karyotyping in myelodysplastic syndromes and related myeloid malignancies. Blood 2011;117:4552–60.

VanGuilder HD, Vrana KE, Freeman WM. Twenty-five years of quantitative PCR for gene expression analysis. Biotechniques 2008;44:619–26.

Voelkerding KV, Dames SA, Durtschi JD. Next-generation sequencing: from basic research to diagnostics. Clin Chem 2009;55:641–658.

Willis AS, van den Veyver I, Eng CM. Multiplex ligation-dependent probe amplification (MLPA) and prenatal diagnosis. Prenat Diagn 2012;32:315–20.

Genetics

Molecular testing is available for many genes involved in inherited diseases. The diseases are often categorized based on their observed inheritance pattern.

- **Autosomal recessive disease** is caused by two mutations on opposite chromosomes with one mutation inherited from each parent. Parents of children with autosomal recessive diseases are unaffected carriers.
- **Autosomal dominant disease** is caused by one mutation in the gene.
- **X-linked recessive disease** is caused by one mutation located in a gene on the X chromosome. Males with a mutation (and only one X chromosome) are affected. Females are typically unaffected carriers (with one mutation, but two X chromosomes), although they may show mild or variable symptoms.
- **X-linked dominant disease** is caused by one mutation located on the X chromosome. Females with a mutation are affected, while it is presumed that one mutation is lethal in males.
- **Mitochondrial disease** is caused by a mutation in the mitochondrial—rather than nuclear—genome. Mutations are inherited through female lines, but both sexes are affected.

This section will describe several common inherited diseases and the molecular testing available. Sample requirements, results, and interpretation will be described for each disease.

Cystic Fibrosis

Cystic fibrosis (CF) is an autosomal recessive disease caused by mutations in the CF transmembrane conductance regulator (*CFTR*) gene. The gene was discovered in 1989; since then, more than 1600 mutations (of varying severity) have been documented. The combinational effect of two severe or moderately severe mutations in this gene affects chloride transport and results in the classic characteristics of CF. The diagnostic test for CF is two positive sweat chloride values (> 60 mmol/L), and it has a clinical sensitivity of approximately 90%.

Characteristics of Classic Cystic Fibrosis

- Chronic sinopulmonary disease
- Gastrointestinal malabsorption
- ± Pancreatic insufficiency
- Meconium ileus in newborns
- Failure to thrive

Nonclassic or atypical CF may present later in life with symptoms often limited to a single organ system. Individuals in this category may have normal or borderline sweat chloride values.

Characteristics of Atypical Cystic Fibrosis

- Idiopathic pancreatitis
- Nasal polyps
- Bronchiectasis or mild pulmonary disease
- Congenital bilateral absences of the vas deferens (CBAVD)
    - CBAVD is the cause of 2%–5% of male infertility. Approximately 75% males have at least one CF mutation, with more than one-half who have two identifiable *CFTR* mutations or one *CFTR* mutation and 5T variant. Others may have one identifiable *CFTR* mutation or two 5T variants.
    - CF testing is recommended for reproductive partners of persons with CBAVD if they are planning to use assisted reproductive technology to have children.

Tests

Several tests are available, from a single known familial mutation to full gene sequencing. Tests are described below.

CF Mutation Panel

- A mutation panel consists of a minimum of the 23 mutations recommended by the American Congress of Obstetricians and Gynecologists (ACOG) and the American College of Medical Genetics (ACMG). Many platforms are available for testing.
  - The recommended mutations were chosen to detect known CF-causing mutations (severe to moderate), with greater than 0.1% of the US population regardless of ethnicity.
  - Other mutations may be added depending on the ethnic makeup of the region the laboratory serves.
  - Reflex tests are recommended for the following:
    - F508C, I507V, and I506V when apparent F508del homozygous results are seen in a healthy adult. These variants may interfere with the assay and give a false-positive result.
    - 5T (in intron 8, also referred to as IVS8) when an R117H mutation is detected. R117H is considered a mild mutation by itself, but has moderate effects when 5T occurs on the same chromosome. If 5T is detected with R117H, family studies or molecular haplotyping can confirm whether the 5T is on the same or opposite chromosome as R117H.
  - If using a standard 23 mutation panel, Table 1 shows detection rates and residual risks after a negative test for different populations.
  - A family history of CF (affected or carrier with unknown mutations) changes the residual risk detailed in Table 1. An individualized residual risk can be calculated by Bayesian analysis for those with a negative result. If familial mutations are tested in the panel, a negative result yields a residual risk equivalent to the population risk.

Table 1. Detection Rates and Residual Risks for Selected Ethnic Groups

Ethnic Group	*Detection Rate (%)*	*Before Test*	*After Negative Test*
Ashkenazi Jews	95	1/25	~1 in 930
European Caucasian	89	1/25	~1 in 220
African-American	65	1/65	~1 in 207
Hispanic-American[a]	73	1/46	~1 in 105[a]
Asian-American[b]	30%[b]	1/90	~1 in 130[b]

[a]This is a pooled set of data and requires additional information to accurately predict risk for specific Hispanic populations.
[b]*Based on F508del only. Other data are limited.*
From Richards CS, Bradley LA, Amos J, et al. Standards and guidelines for CFTR mutation testing. Genet Med 2002;4:379–91.

CFTR Gene Sequencing

- Sequencing of the entire coding region of the gene, plus intron/exon boundaries is widely available. Deep intronic mutations may be included (such as the 3849 +10kBC>T mutation that is common in some populations). The gene promoter may also be sequenced, although few mutations have been described in this region.

CFTR Deletion/Duplication Analysis

- Up to 10% of *CFTR* mutations may be deletions. Several techniques are available, but two that are typically used are quantitative multiplexed PCR or multiplex ligation-dependent probe amplification (MLPA). A recurrent deletion of exons 2–3 has been reported in eastern European populations.

Specific Mutation Testing

- If the family-specific mutation is known, family members can be tested for the specific mutation. If the familial mutation is a part of the panel, the panel is typically a less expensive test. If the familial mutation is not part of the panel, it can be tested by sequencing the exon where the mutation located.

Indications for Testing

The following are indications for CF testing.

- Individuals affected with CF (positive sweat chloride of > 60 mmol/L)
- Reproductive partner of a person affected with or a carrier of CF
- Relative(s) of CF patients or known carriers
- Caucasian and Jewish persons who are pregnant or planning a pregnancy, although testing is made available to other ethnic groups

Figure 1 shows a strategy for molecular confirmation of cystic fibrosis.

The following are testing indications for atypical CF:

- Individuals with borderline sweat chloride
- Males with CBAVD
- Individuals with pancreatitis (test also genes known for hereditary pancreatitis, *PRSS1*, and *SPINK1*)
- Individuals with bronchiectasis

Figure 2 shows a strategy for molecular confirmation for atypical CF-related disorders.

Clinical scenarios determine which testing protocol is most appropriate.

- CF mutation panels are used for the following:
    - Newborn screening positive for immunoreactive trypsinogen (IRT)
    - Population screen for carrier testing

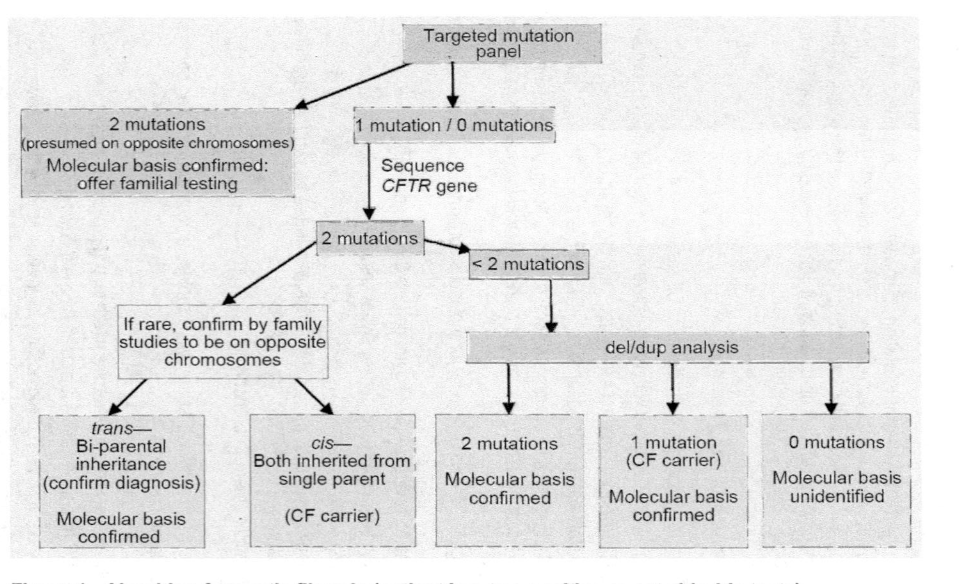

Figure 1. Algorithm for cystic fibrosis (patient has two positive sweat chloride tests).

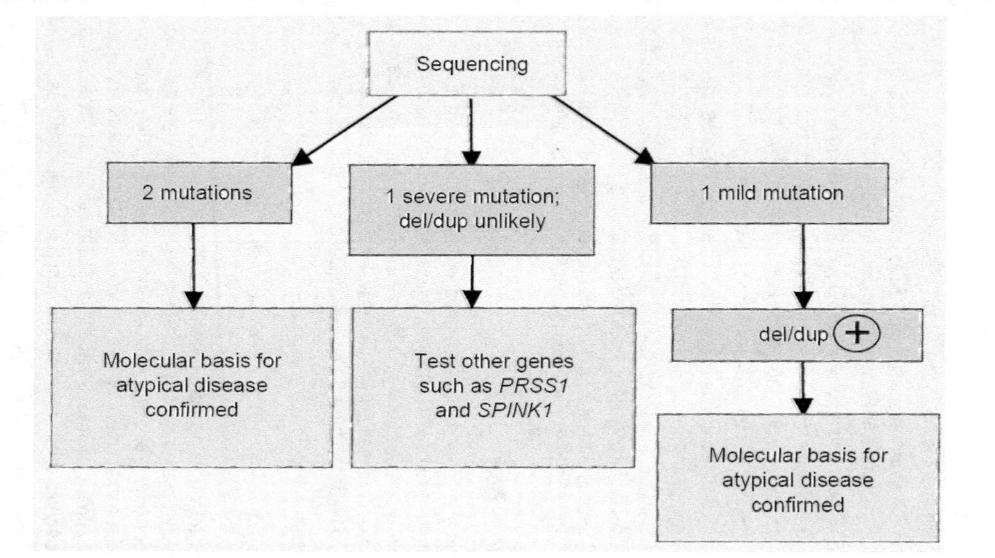

Figure 2. Algorithm for testing for monosymptomatic disease.

  - Infants with failure to thrive
  - First tier testing for individuals affected with CF
- The CF mutation panel +IVS8 5T is used as a first tier test for individuals affected with atypical CF.
- *CFTR* gene sequencing is used under the following situations:
  - Second tier testing for affected individuals (classical and atypical CF) in which panel testing did not identify two mutations
  - First tier test for individuals suspected of atypical CF
- Deletion/duplication testing is used if other testing does not explain the symptoms present in an individual:
  - Third tier test for classically affected individuals if fewer than two mutations are detected by sequencing
  - Second tier test for individuals suspected of being affected with atypical CF if only one mild mutation is found by sequencing.

Sample Type

DNA from whole blood (ethylenediaminetetraacetic acid [EDTA] or acid-citrate-dextrose [ACD]), blood spots, saliva, cheek swabs, amniotic fluid, or amniocytes yield adequate DNA for mutation panels or for family specific mutations. Whole blood samples or cultured amniocytes are recommended for full gene sequencing and deletion/duplication analysis.

Results

- **Negative:** A "negative" result for a mutation panel is more accurately phrased as "None of the tested mutations were detected." For full gene or deletion/duplication testing, a negative result indicates that "no mutation was detected." For family specific results, a negative result means that "the familial mutation was not detected."
- **Positive:** When at least one mutation is detected, results are considered "positive." The mutation is listed by either the DNA mutation or the predicted protein change. Standard nomenclature

is encouraged, but since literature references often use 'legacy' names, the older nomenclature may also be included in reports. Mutations detected are listed as "heterozygous" (one copy) or "homozygous (two copies).

Interpretation

The interpretation of the genotype depends on the indication for testing. If no mutations are detected, the residual risk of being affected or a carrier can be reduced, but not eliminated. If patients are diagnosed with CF by clinical symptoms and sweat chloride testing, they are considered to be affected whether or not two mutations can be identified. The following describes the interpretation for positive results.

- If one mutation is detected and the patient is unaffected, the individual is a carrier.
- If one mutation is detected and the patient is affected or the affected status is unknown, the genotype is interpreted as "this individual is at least a carrier and may be affected if a mutation is present that is not detected by the assay."
- If two mutations are detected and have not been described on the same chromosome or confirmed to be on opposite chromosomes by family studies, the individual is predicted to be affected. Whether they are affected with classical CF or atypical CF depends on the severity of each mutation and in combination.
- If two mutations are detected that have been described together on the same chromosome or confirmed by familial studies (complex mutation), they should be considered as a single mutation. If no other mutation is found, an individual with a complex mutation is considered a CF carrier.
- Although many CF variants are characterized, variants with uncertain clinical significance may be detected. Until further characterized, these variants cannot be considered CF mutations.

Prediction of severity of disease is possible when well-characterized mutations are detected. The following describes the predicted phenotype when different combinations of mutations are identified.

- Two severe mutations: Classic CF, pancreatic insufficient
- Two moderate mutations: Classic CF, but likely pancreatic sufficient
- One severe/one moderate mutation: Classic CF ± pancreatic sufficiency
- One severe or moderate/one mild mutation: Atypical CF
- Two mild mutations: Often asymptomatic, may have variable atypical and mild symptoms.

Limitations

If familial mutations are not part of the panel, targeted sequencing will need to be performed. Rare variants with primer or probe sites may interfere with the assays.

Comments

- Including mild mutations in a mutation panel is discouraged if the panels are used for population carrier or newborn screening, since prediction of phenotypes is unreliable.
- To properly interpret results, laboratories need information as to whether the individual has CF (or CF-related) symptoms, if a family history is present and the ethnicity.
- If one severe mutation is identified in a patient with atypical CF, deletion/duplication testing is unlikely to identify a second mutation. Since deletions and duplications are typically severe mutations, such an individual would be expected to have symptoms consistent with classic CF.

Thrombophilia

Thrombophilia is common malady, with 2 million cases of deep venous thrombosis (DVT) per year in the U.S alone. Both acquired and inherited risk factors play a role in the development of this condition.

Characteristics

- Complications include pulmonary embolus
- Post-thrombotic syndrome is present in 60% or the cases and is characterized by:
  - Chronic vessel obstruction and valve destruction by an organized thrombus
  - Venous hypertension
  - Chronic edema, pain, leg ulcers

Acquired Factors

- Pregnancy/puerperium
- Hormone therapy
- Surgery
- Immobilization
- Malignancy
- Chemotherapy
- Previous DVT or superficial venous thrombosis
- Heavy smoking
- Obesity
- Lupus anticoagulant

Inherited Factors

These include mutations in procoagulant proteins or major anticoagulant proteins. Further details of inherited factors are discussed below.

- **Mutations in procoagulant proteins**
  - Factor V
    - 1691G>A substitution, R506Q, also known as factor V Leiden (FVL).
      - FVL is associated with activated protein C (APC) resistance, by preventing inactivation of factor Va by APC at one of three APC cleavage sites.
      - It accounts for up to 95% of APC resistance
      - It results in a continual mild hypercoaguable state
    - Factor V R2 haplotype
      - The R2 haplotype is a combination of nine variants in exon 13 of the *F5* gene that exist on the same chromosome. It has always been observed on the opposite chromosome as FVL
      - It increases risk for thrombotic event in individuals heterozygous for FVL
      - Testing for one of the variants in the haplotype reveals the presence of one of the polymorphisms in the haplotype
    - Other factor V mutations have been described. Their characterization to date suggest a mild or no effect on APC resistance
      - FV Cambridge (p.Arg306Thr)
      - FV Hong Kong (p.Arg306Gly)
      - FV Liverpool (p.Ile359Thr)
      - Asp79His (p.Asp79His)
  - Factor II (prothrombin)
    - 20210G>A substitution
      - The mutation is located in the 3' untranslated region that affects the polyadenylation signal that is important for mRNA stability
      - It results in increased mRNA expression and increased Factor II levels (30% in heterozygotes, 70% in homozygotes)

- **Mutations in major anticoagulant proteins**. Since these are rare and without common mutations, they are not routine molecular tests.
  - Protein C
  - Protein S
  - Antithrombin III
- **5,10-Methylenetetrahydrofolate reductase (MTHFR; 677C>T and 1298A>C)**. Hyperhomocysteinemia has been associated with venous thrombosis and cardiovascular disease, although the underlying mechanisms are unclear. There are many causes for hyperhomocysteinemia such as renal failure, hypothyroidism, leukemia, psoriasis, and drug therapy. The MTHFR enzyme is involved with folate metabolism and converts 5,10-methylenetetrahydrofolate to 5-methyltetrahydrofolate, which is the primary circulatory form of folate. Variants in MTHFR have been associated with increased homocysteine, although environmental factors also play a major role. Two variants in the *MTHFR* gene have been described.
  - 677C>T (p.665C>T, p.Ala222Val) is an autosomal recessive thermolabile mutation, requiring homozygosity for mild increase of total plasma homocysteine.
    - It is common in the population; 30%–40% are heterozygous and 10%–15% are homozygous
    - It accounts for one-third of hyperhomocysteinemia cases
  - 1298A>C (p.1286A>C, p.Glu429Ala) is only associated with increased homocysteine when seen in combination with 677C>T. Homozygosity for 1298A>C is not associated with increased homocysteine.

Tests

Common variants for each gene may be tested separately or as a panel. Multiple platforms are available. Other tests are important in the workup of thrombophilia and include plasma homocysteine, prothrombin time and APC resistance. However, molecular testing is necessary to distinguish FVL heterozygosity from homozygosity.

Indications for Testing

An ACMG/College of American Pathologists (CAP) consensus statement in 2001 gives testing indications for FVL. Testing is indicated for individuals after a venous thrombosis occurring:

- At < 50 y
- During pregnancy or the puerperium
- With oral contraceptives or hormone replacement therapy
- At any age if strong family history
- In unusual sites unprovoked, at any age
- Recurrently at any age
- Smokers < 50 y with one provoked venous thrombosis

Other testing indications include:

- Unexplained second or third trimester pregnancy loss
- Unexplained severe preeclampsia, placental abruption or intra-uterine growth restriction
- Testing of relatives of patients with FVL

Similar criteria may be used for factor II (prothrombin). However, a recent ACMG practice guideline recommends against ordering MTHFR mutations as part of a routine evaluation for thrombophilia.

Sample Type

DNA from whole blood (EDTA or ACD), saliva, or cheek swabs, yield adequate DNA for single mutations or thrombophilic targeted mutation panels.

Results

Results are given for each variant as Negative (not detected), heterozygous (one copy) or homozygous (two copies). If FVL is determined to be heterozygous, R2 haplotype testing could be considered.

Interpretation

One copy of a single mutation (FVL or factor II) will increase the risk for thrombophilia. Two copies of the same mutation or two mutations in combination will increase the risk further. For hyperhomocysteine and thrombotic risk due to MTHFR, homozygosity for 677C>T or compound heterozygosity for 677C>T and 1298A>C mutations in the *MTHFR* gene are needed for increased homocysteine. Although the immediate therapeutic approach may not change, considerations may be given for surgery or prolonged immobilization

Relative risks of factor V and prothrombin genotypes are shown in Table 2.

Inherited factors can combine with environmental factors to increase risk over either alone. Table 3 shows the combined risk of oral contraceptives and FVL.

Table 2. Relative Risks for Factor V and Prothrombin Genotypes

Genotype	*Relative Risk of Venous Thrombosis*
FVL (heterozygous)	4–8
FVL (homozygous)	80
FII 20210G>A (heterozygous)	2–5
FII 20210G>A (homozygous)	Increased, but risk unknown
FVL (heterozygous) + 20210G>A (heterozygous)	20
FVL + R2 haplotype	Increased over FVL alone

FVL, factor V Leiden.

Table 3. Risk Factors for Oral Contraceptives and Factor V Leiden

Risk Factor	*Relative Risk of Venous Thrombosis*
Oral contraceptives	4
Oral contraceptives and FVL (heterozygous)	35
Oral contraceptives and FVL (homozygous)	100

FVL, factor V Leiden.

Limitations

Other mutations are typically not tested. Rare sequence variants within primer and probe sequences may interfere with testing.

Comments

Other mutations may be detected coincidently with the targeted mutations. One example is the prothrombin 20209C>T variant. This has been seen in African-Americans, but its role in thrombophilia has not been confirmed.

Fragile X Syndrome

Fragile X syndrome is the most common heritable form of intellectual disability and found in all populations. It is considered an X-linked dominant disorder with reduced penetrance in females. The cause of Fragile X syndrome is large expansions (full mutation) of a 5' UTR region of CGG trinucleotide repeats leading to hypermethylation and inhibition of gene transcription. In rare cases, a gene deletion or point mutations inactivating the gene can cause the disease. The incidence of disease is 1:4000 males and 1:8000 females, with a carrier frequency estimated to be 1.3%.

Characteristics

- Intellectual disability: Moderate in males, mild in females
- Behavioral hyperactivity, preservative speech, social anxiety, poor eye contact, hand flapping or biting, autism spectrum disorder
- Hyperflexibility
- Physical manifestations: Macroorchidism, long narrow face, prominent ears and jaw

Premutation carriers are also at risk for later-onset disease, presumably by a different mechanism due to messenger RNA (mRNA) accumulation. Women are at risk for premature ovarian insufficiency (POI) while males (and rarely women) are at risk for Fragile X–related tremor/ataxia (FXTAS).

Characteristics of Fragile X–Related Tremor/Ataxia

- Late-onset, progressive
- Cerebellar ataxia/intention tremor
- Short-term memory loss,
- Executive function deficits
- Cognitive decline
- Lower-limb proximal muscle weakness, and autonomic dysfunction

The categories of FX alleles with their sizes and methylation are shown in Table 4.

Tests

Polymerase chain reaction (PCR) and Southern blot are performed sequentially (PCR reflexed to Southern analysis) or concurrently. PCR sizes normal, intermediate and smaller premutation alleles. Southern analysis sizes large premutations and full mutations and determines methylation status of expanded alleles.

Table 4. Sizes and Methylation Statuses of Fragile X Alleles

Categories	*Size of CGG Repeats*	*Methylation Status*
Full mutation	>200–230	Methylated
Premutation	56–200	Unmethylated[a]
Intermediate	45–55	Unmethylated
Normal	5–44	Unmethylated

[a]Unmethylated alleles > 200 repeats may also be considered premutation alleles.

Indications for Testing

Carrier testing for premutations and diagnostic testing for full mutations are available. Indications for each are listed below:

- Fragile X diagnostic testing is used for individuals with unexplained intellectual disability, developmental delay or autism, especially for intellectual disability with physical or behavioral characteristics or family history consistent with Fragile X syndrome. Fetal predictive testing is available for women who are known premutation carriers.
- Premutation carrier testing is used for individuals with a family history of Fragile X syndrome or X-linked intellectual disability. In addition, females with POI or males (or rarely females) with late-onset cerebellar ataxia and intention tremor are tested for premutations.
- No recommendations have been made for population screening, either carrier (premutation screening) or newborn (full mutation or methylation screening).

Sample Type

DNA from whole blood (EDTA or ACD), blood spots, saliva, cheek swabs, amniotic fluid, or cultured amniocytes yield adequate DNA for

PCR analysis. Southern blot requires DNA from whole blood (EDTA or ACD) or cultured amniocytes.

Results

Results are given as normal, intermediate, premutation, full mutations or mosaics (premutation/full mutation or rarely normal/full mutation). Methylation status is also given for expanded alleles.

Interpretation

Interpretations for each result are shown below:

- **Normal:** 5–44 repeats, normal methylation patterns: Rules out diagnosis of Fragile X syndrome or premutation carrier status, as well as Fragile X–related POI or FXTAS.
- **Intermediate:** 45–54 repeats, normal methylation patterns: Rules out Fragile X syndrome and premutation carrier status. Individuals with intermediate alleles are not affected, but the alleles are unstable, and could eventually expand to a premutation. Although intermediate alleles are generally not considered risk alleles for POI and FXTAS, studies to confirm or disprove this are needed.
- **Premutation:** 55–200 repeats, normal methylation patterns; confirms carrier status in males and females.
    - In females, premutations confer a risk for expansion in the next generation resulting in children affected by Fragile X. The risk of expanding into a full mutation in offspring is dependent on the size of the repeat. Premutation females are at risk for POI or rarely ataxia. Table 5 shows the risk of a premutation allele expanding if it is transmitted to offspring.
    - Premutation alleles inherited through fathers will be passed to all of their daughters, but are unlikely to expand to full mutations in one generation. Premutation males are at risk for FXTAS. Males do not pass premutations to sons.

Table 5. Risks of Premutation Allele Expansion

Number of Maternal Premutation CGG Repeats	*Approximate % Risk of Allele Expanding to a Full Mutation (%)*
56–59	14
60–69	20
70–79	58
80–89	72
90–99	94
> 100	100

Adapted originally from Warren & Nelson 1994; modified according to Nolin et al. 1996. FMR-1 related disorders. GENEReviews at www.genetests.org.

- **Full mutation:** > 200–230 repeats, methylated; Confirms diagnosis of Fragile X syndrome, as the gene is methylated and predicted to be inactive.
  - Females: May be mildly affected, or normal, depending on X inactivation of normal alleles. Females with full mutations have 50% risk of offspring (regardless of sex) with full mutations.
  - Males with full mutations are affected with Fragile X.
- **Mosaic:** Confirms a diagnosis of Fragile X syndrome in males. Both premutation (unmethylated) and full mutations (methylated) are present. Severity of symptoms cannot be predicted by a mosaic pattern, but may be milder than in nonmosaic full mutations due to unmethylated alleles.

Limitations

Comparison of limitations of PCR and Southern analysis are listed below.

- PCR has difficulty amplifying through large CGG repeats, although amplification into full mutation range is possible.
  - In females, preferential amplification of normal alleles makes it difficult to distinguish between normal homozygous alleles and one allele/one undetected expanded allele.
  - In males, normal/full mosaics (most likely due to contraction into the normal range) occur in about 3% of full mutations, and pose the same difficulty as apparent homozygous females.
  - If expanded alleles are reflexed to Southern analysis, apparent homozygous females will need to be reflexed, while normal/full mutation males will be missed.
- Southern blot analysis can detect premutations from 80 repeats to large full mutations, although precise sizing is difficult by Southern analysis.
- Fragile X due to point mutations will not be detected by either method and requires gene sequencing. Other forms of Fragile X such as Fragile X E will not be detected.

Comments

- In males a complete gene deletion will be suspected if no bands can be seen on Southern analysis.
- Efforts have been made to validate dosage (confirm homozygosity) in females. However, this has not been implemented into routine practice.
- A recently available triplet-primed PCR detects "expanded alleles" and resolves the difficulty of PCR-only results for apparent homozygous females and mosaic normal/full mutation males.
- Methylation-specific PCR has also been described to determine methylation status from PCR alone or in combination with PCR product blotting. Its use is as follows:
  - Detect methylated alleles in males
  - Determine whether borderline pre/full expansions are methylated
  - Methylation and size simultaneously determined

Hemophilia A

Hemophilia A is an X-linked recessive disease that can range from mild to severe. The prevalence of disease is approximately 1:4000–5000 male births. Diagnosis is made by a factor VIII activity level of less than 50%, with a normal von Willebrand factor level. Ten percent of carrier females are mildly affected. It is an X-linked recessive disorder caused by mutations in the factor 8 (*F8*) gene. Severity of disease is correlated to the type of mutation present. Two known inversions (involving intron 22A and intron 1) are severe mutations. Missense mutations are more likely to be mild mutations. Carrier or prenatal testing is best confirmed by molecular testing.

Characteristics

- Severe disease is characterized by spontaneous joint to deep-tissue bleeding. Factor VIII activity is < 1%, and is usually diagnosed early in life. Treatment is prophylactic infusions of factor VIII concentrate. Approximately 50% of patients are severely affected.
- Moderate disease has a factor VIII activity of 1%–5%. Approximately 10% of patients are moderately affected.
- Mild disease shows and factor VIII activity of 6%–35%. Individuals with mild disease have prolonged or recurrent bleeding after surgery or injury. Approximately 40% of patients are mildly affected.

Tests

Several tests listed below are needed for a complete evaluation for *F8* mutations.

- Inversion tests include the common intron 22A and rare intron 1 inversions. The test utilizes a restriction digest, ligation, and PCR. Inversion testing should be the first test for patients with severe hemophilia.

- Sequencing is performed for severe cases when an inversion is not present. It is a first test for patients with mild hemophilia.
- Deletion/duplication testing is performed when a mutation is not found by inversion testing or sequencing.
- The family-specific mutation can be tested, when known.

Indications for Testing

The three common reasons for molecular testing for *F8* mutations are:

- Confirming the mutation in affected individuals
- Determining carrier status of females with a family history
- Fetal testing for a family specific mutation

A testing algorithm is shown in Figure 3.

Sample Type

DNA from whole blood (EDTA or ACD), saliva, amniotic fluid, or amniocytes yield adequate DNA for family specific sequence mutations. Whole blood samples or cultured amniocytes are recommended for inversion testing, full gene sequencing, or deletion duplication analysis.

Results

Results are given as a mutation (or variant) detected or no mutation detected. Although rare, females may have a mutation on each X chromosome

Interpretation

The presence of a pathogenic mutation identified in any of the tests confirms either affected status (in males) or carrier status (in females). Although the severity of disease is determined by the factor 8 activity level, mutations can also be classified as "mild," "moderate," or "severe" based on activity levels.

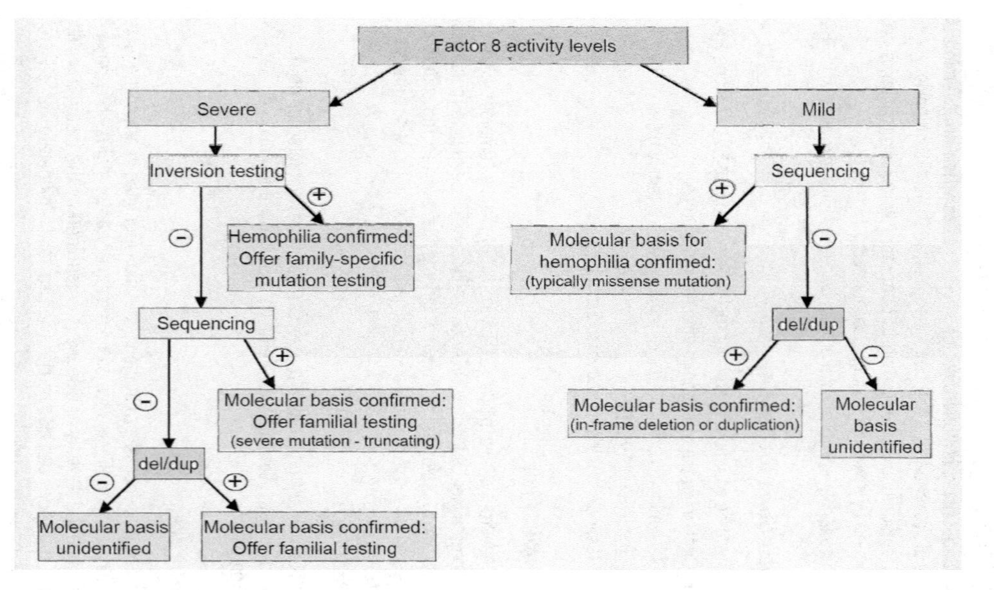

Figure 3. Algorithm for hemophilia.

Limitations

For sequencing assays, mutations in untranslated regions may not be interrogated. It is important to rule out von Willebrand disease (vWD) or hemophilia B because they are caused by mutations in different genes.

Hereditary Hemochromatosis

Hereditary hemochromatosis (HH) is characterized as an iron-overload disease. It is an autosomal recessive disease caused by mutations in the *HFE* gene. One common mutation (p.Cys282Tyr or C282Y, c.845G>A) and two low penetrance variants (p.His63Asp or H63D, c.187C>G and p.Ser65Cys or S65C, c.193A>T) have been described.

Tests

Most tests are targeted for the common mutations rather than gene sequencing. All laboratories test for the C282Y mutation, while some laboratories include H63D and S65C. The mutations may be tested together in a multiplex PCR or singly.

Indications for Testing

Testing is available for individuals with high serum ferritin or transferrin saturation or individuals with a family history of hemochromatosis or iron overload.

Sample Type

DNA from whole blood (EDTA or ACD), saliva, and cheek swabs, yield adequate DNA for *HFE* mutation panels.

Results

Results are given as negative (none of the targeted mutations detected) or heterozygous or homozygous for each of the mutations

Interpretation

Homozygosity of C282Y is confirmatory or predictive of HH, although penetrance is reduced. H63D/C282Y or S65C/C282Y compound heterozygosity is also associated with HH, but with a lower penetrance than C282Y homozygosity. H63D or S65C homozygosity is rarely associated with HH, and is not considered diagnostic.

Limitations

Juvenile HH is due to another gene.

Comments

As HH is an adult onset disorder, testing for asymptomatic juveniles is not recommended.

Rett Syndrome

Rett syndrome is a form of intellectual disability predominantly affecting females. It was considered X-linked dominant, being lethal in males, until gene duplications were discovered in males with severe disability. The prevalence is 1:10,000 females, with males still being ascertained. The cause is mutations in the *MECP2* gene, whose product regulates genes involved in neuronal maturation. Early development of affected females is normal, with a loss of motor skills and communication. There is a high variability in disease severity and progression. Mutations in this gene are identified in 90%–95% classic Rett syndrome. However, *MECP2* mutations can result in nonclassic phenotypes.

The *MECP2* gene has many isoforms. Most are differences in 3′ untranslated region (UTR) lengths. However, two main isoforms exist with differing exons: *MECP2B* and *MECP2A*.

- *MECP2B* (brain, thymus, lung) includes exons 1,3,4
- *MECP2A* includes exons 2,3,4

Characteristics

- Severe intellectual disability, with progressive neurological findings, hypotonia, microcephaly, dysmorphic features (e.g., large ears, hypospadias, flat nasal bridge, autistic-like features)
- Duplications and a triplication have been reported
  - 0.4- to 0.8-Mb duplications (10 different cases)
  - 0.2- to 2.2-Mb duplications (including *L1CAM* gene)

Tests

Sequencing of exons 1–4 will detect most known sequence variants. Deletion/duplication (MLPA) testing can be reflexed or tested concurrently.

Indications for Testing

MECP2 testing is performed for females with Rett-like symptoms and in males with severe mental retardation.

Sample Type

DNA from whole blood (EDTA or ACD) is recommended for full gene sequencing and deletion duplication analysis. When parental studies are needed to confirm a de novo mutation, DNA from whole blood (EDTA or ACD) or saliva yields adequate DNA.

Results

Results are given as positive for a mutation (variant) or no mutation detected.

Interpretation

Identification of a sequence variant is followed by testing parents to confirm a *de novo* mutation. If it is *de novo*, it is likely to be causative of Rett syndrome. If it is seen in an asymptomatic mother, it is less likely to be causative (unless evidence of skewed lyonization exists). If it is found in the father, this is strong evidence that the variant is benign. A partial or full gene deletion or duplication confirms the diagnosis in females. Duplications and rare point mutations confirm the diagnosis in males.

Limitations

- Variants are often family-specific and may not be well characterized; therefore, a result may be of a variant of uncertain significance.
- Deletions/duplication breakpoints may not be determined.
- Deletion or duplication at the 3′ end of the gene may extend into other genes that are not tested in the assay, but could contribute to the phenotype.

Comments

Alternative transcripts are seen with differential expression in brain and blood. Variants in exon 1 may be expressed in the brain, but not the blood.

Hemoglobinopathies

Disorders of hemoglobin are among the most common heritable condition worldwide. The beta gene is located on chromosome 11p15.5, while the alpha gene cluster is on chromosome 16p13.3. Genes are differentially expressed during development.

- Hemoglobins are designated as follows:
  - Embryo: z_2e_2(Hb Gower); z_2g_2(Hb Portland); a_2e_2(Hb Gower 2)
  - Fetus: a_2g_2(Hb F)
  - Adult: a_2b_2(Hb A); a_2d_2(Hb A_2)

More than 1200 mutations have been described in the alpha (*HBA*) and beta (*HBB*) globin genes. Variants were initially differentiated by electrophoretic mobility.

Whereas gene deletions are the most common *HBA* mutations, sequence variants are more common in *HBB*. Hemoglobin disorders are classified as hemoglobinopathies or thalassemias. Both alpha and beta thalassemias are generally recessive inheritance, although a single copy of a severe mutation can result in mild symptoms such as mild microcytic anemia. Assays can be designed for known *HBB* deletions (such as the 619bp deletion found in Indian and other Asian populations) and fusion genes (such as Hb Lepore).

- **Hemoglobinopathies**: Mutations in the globin genes resulting in the synthesis of structurally abnormal globin subunits.
- **Thalassemias**: Reduced synthesis of structurally normal globin subunits resulting in unbalanced synthesis of beta and alpha globin chains. Excess chains are toxic to developing erythrocytes.

Disorders of beta and alpha globin are considered separately.

- **Beta globin**
  - Hemoglobinopathies are due to missense mutations that change hemoglobin (Hb) function. The most common mutations are Hb S, C, and E. Such mutations are often identified by Hb high performance liquid chromatography (HPLC).
  - Hb S (Glu6Val, c.20A>T): Homozygosity of Hb S is the molecular basis of sickle cell anemia. Sickle cell anemia has a prevalence of about 1:500 among African-Americans, with about 10% of this population being carriers (designated sickle cell trait).

    - Compound heterozygosity with other *HBB* mutations can result in sickle cell disease.
    - Characteristics of sickle cell disease include: vascular occlusion, episodic pain crises, pain and swelling of hands and feet, stroke, hemolytic anemia and infection.
  - Hb C (Glu6Lys, c.19G>A): The Hb C allele is significant when inherited with Hb S as a compound C/S and can result in sickle cell disease. Homozygous Hb CC individuals have mild hemolytic anemia. The Hb C allele is carried by approximately 2% of African-Americans.
  - Hb E (Glu26Lys, c.79G>A): Hb E is not strictly a structural mutation, since phenotypes are dependent on a second mutation. Hb E is considered a mild mutation, as homozygosity for Hb E results in mild hemolytic anemia. However, when inherited as a compound heterozygous with a beta thalassemia mutation, the resulting phenotype is beta thalassemia of varying severity.
  - Beta thalassemias are classified into beta (0), or beta (+) thalassemias depending on the amount of beta chain synthesized.
    - Beta (0) indicates that very little to no beta chain is produced.
    - Beta (+) indicates that some beta chain is produced, but in reduced quantities. Beta (+) mutations with significant reduction of beta chain synthesis are more severe than beta (+) mutations with some reduction of beta chain synthesis.
    - Individuals with beta thalassemia major have two beta thalassemia mutations.
    - Carriers of a single beta thalassemia mutation are said to have beta thalassemia minor.
    - Characteristics of beta thalassemia major include severe microcytic anemia, with iron overload and heart disease as a significant complications. Patients are treated with regular transfusions and, if needed, a chelating agent for iron overload.

- **Alpha globin proteins**
  - Coded by two genes per chromosome 16 for a total of four genes per person. Therefore, the normal configuration is:
    $\alpha\alpha/\alpha\alpha$ (or for laboratory information systems: aa/aa)
  - Deletions of one or both copies on a chromosome make up as much as 95% of alpha globin deletions, with 29 known deletions. More than 100 known small deletions or mutations have also been reported.
    - Carrier rates for alpha globin deletions are ethnic-specific:
      - African descent: ~30%
      - Mediterranean descent: 3–30%
      - Asian descent: 5–40%
      - Northern European: 0.1%
      - Southern European: 1%–19%
    - Common deletions are often performed in a single test. Deletions include $-a^{3.7}$, $-a^{4.2}$, $--^{SEA}$, $--^{MED}$, $--^{FIL}$, $-(a)^{20.5}$, $--^{THAI}$.
    - Hb H disease (–/-a) is characterized by beta chain tetramers [b^4].
    - Hb Bart's/hydrops fetalis (--/--) is characterized by gamma chain tetramers [g^4].
  - Alpha thalassemia point mutations
    - About 5% of alpha thalassemias are due to point mutations. These mutations often cause more severe thalassemia than single gene deletions.
    - The most common point mutation is Hemoglobin Constant Spring, in which the stop codon is mutated, resulting in unstable RNA, and low production levels of a large alpha globin protein.

Tests

Testing for hemoglobin disorders often begin with HPLC analysis, isoelectrofocusing, or electrophoresis, although these analyses identify structural variants better than thalassemias unless beta4 (Hb H) or gamma4 (Hb Barts) is present. Once an abnormal hemoglobin is detected, molecular testing can further characterize the mutation.

- **Beta globin**
  - Tests for sickle cell anemia include a targeted mutation analysis for Hb S and Hb C.
  - Further evaluation is possible through *HBB* gene sequencing.
    - The coding region and intron/exon boundaries of the three exons are interrogated.
    - Since known pathogenic mutations exist deep within introns and the 5′ promoter and 3′ UTR, complete beta globin gene sequencing will include targeted testing for known deep intronic mutations and the 5′ and 3′ UTR.
  - Known deletions of the beta globin gene and can be tested by amplifying over the deletion breakpoints or by a general deletion/duplication test method, such as MLPA or exonic level microarray.
- **Alpha globin**
  - A common deletion mutation panel is available.
  - Gene sequencing includes both the *HBA1* and *HBA2* genes.

Indications for Testing

Molecular testing for hemoglobin related disorders are useful to confirm a diagnosis for symptomatic individuals, to provide carrier testing in individuals with a family history or to clarify an abnormal HPLC or gel electrophoresis pattern. Prenatal diagnosis is available when both parents are known carriers. Population screening for alpha thalassemia deletions is available for healthy individuals of African, Mediterranean, and Southeast Asian descent.

Sample Type

DNA from whole blood (EDTA or ACD), blood spots, saliva, cheek swabs, amniotic fluid or amniocytes yield adequate DNA for mutation panels or for family specific mutations. Whole blood samples or cultured amniocytes are recommended got full gene sequencing and deletion/duplication analysis.

Results

A positive result is the detection of one mutation (for a carrier) or two mutations (for affected individuals) by any of the methods described. For sequence variants, common nomenclature or standard nomenclature or variant names (such as Hb Baltimore) are used. The first amino acid is removed in the processed beta globin protein, so the common name for the protein will be one amino acid different than in standard nomenclature. Deletions in the alpha globin genes are reported as their common names (e.g., SEA), and as heterozygous, homozygous or compound heterozygous of two deletions.

Interpretation

- **Beta globin:** Interpretation for beta globin variants depends on the combination that is seen. Examples of *HBB* sequence interpretations are:
    - For structural variants:
        - Heterozygosity for Hb S predicts a clinical presentation of sickle cell trait.
        - Homozygosity for Hb S predicts a clinical presentation of sickle cell disease.
        - The combination of two mutations can cause severe sickle cell anemia, if the mutations are on different chromosomes. Some mutations have been reported on the same chromosome. Comparing to HPLC results or testing parental samples can confirm whether the variants are on the same of different chromosomes.
    - For thalassemia variants:
        - One copy of a pathogenic mutation is consistent with beta thalassemia minor
        - Two pathogenic mutations frequently result in beta thalassemia major, although the clinical presentation may vary due to other genetic modifiers.

- For structural and thalassemia compound mutations
  - One copy of the Hb S mutation and one copy of a beta (+) thalassemia mutation is consistent with Hb S/beta (+) thalassemia, a form of sickle cell disease that presents with varying clinical severity.

- **Alpha globin**
  - Deletion results are interpreted as follows:
    - Normal (aa/aa)
    - Silent carrier (-a/aa, a^{T}a/aa)
    - Carriers with mild hematologic changes (anemia, reduced cell hemoglobin and mean cell volume values) (-a/-a,--/aa)
  - The two different types of carriers are important to distinguish for reproductive purposes. Table 6 shows the possible offspring genotypes for couples with alpha globin deletion results.
  - The interpretation is dependent on the combination of variants detected. Examples of *HBA* sequence interpretations are below.
    - Homozygosity for Hb CS has previously been reported to result in hematologic findings that range from mild or moderate anemia to Hb H-like disease.

Table 6. Possible Offspring Genotype for Couples With Alpha Globin Deletion

Parent 1 Genotype	*Parent 2 Genotype*	*Possible Offspring Genotypes*
-a/-a	-a/-a	-a/-a (100%)
-a/-a	--/aa	-a/aa (50%) or –a/-- (50%)
--/aa	--/aa	aa/aa (25%), aa/-- (50%) or --/-- (25%)

    - The combination of two pathogenic point mutations is predicted to result in a clinical presentation that resembles Hb H disease.
    - Heterozygous carriers of *HBA* pathogenic point mutations are reportedly normal. If the variant has not been reported to be associated with a clinical phenotype, it is likely benign.
  - Alpha globin deletion analysis and sequence analysis may need to be interpreted together. An example of a combined interpretation is below.
    - Individuals carrying a mutation in *HBA2* and the -3.7 Kb deletion are predicted to have mild anemia and microcytosis consistent with alpha(0) thalassemia (trait).

Limitations

Sequencing analysis will not detect large gene deletions. If a patient has one copy of a gene deletion, sequencing analysis will not detect it, and will appear normal. If a patient has two copies of a gene deletion, sequencing will not be able to yield a result.

Comments

Unless family specific mutations are known, or HPLC or other testing indicates an alpha globin point mutation, alpha globin deletion testing should be performed first. For carrier testing of alpha thalassemia, the common deletions should be tested first.

Huntington Disease

Huntington disease (HD) is a progressive neurodegenerative disorder characterized by cognitive, motor and psychiatric disturbances. It is inherited in an autosomal dominant manner, with high penetrance. It is due to a trinucleotide repeat expansion (CAG or polyGlutamine tract) within the coding region of the *HTT* gene, which produces the huntingtin protein. *HTT* is widely expressed in neuronal and non-neuronal tissues. Onset roughly corresponds to repeat size. Incidence of disease is 1:10,000, with typically a midlife onset, although up to 5% of the cases are juvenile. The progressive problems with coordination, judgment, and thinking result from selective neuronal loss from caudate nucleus and putamen. Death typically occurs 15–20 y after symptoms begin. It is hypothesized that an HD mutation leads to toxic gain-of-function in the huntingtin protein.

Characteristics

Symptoms include psychiatric, motor, and cognitive decline, as described below.

- Psychiatric disturbances: Depression, increased suicide risk, personality and mood changes, irritability, aggressiveness, apathetic demeanor, paranoia, and delusions
- Motor symptoms: Chorea, gait, fine motor control, hyperreflexia, bradykinesia, rigidity, dysarthria, dysphagia
- Cognitive disturbances: Attention and concentration problems, memory deficits, slower thought processes, visuospatial difficulties, word-finding difficulties, lack of awareness

Although there is no cure for HD or treatment that can slow progression, some symptoms may be alleviated by medication.

- Psychiatric disturbances
  - Psychotropic medications (selective serotonin reuptake inhibitor [SSRI])
- Motor symptoms
  - Neuroleptics for chorea
  - Anti-Parkinsonian agents for hypokinesia
  - Prone to adverse events
- Cognitive impairment: No treatment
- Movement training or speech therapy

The juvenile form of HD is defined as age of onset less than 20 y, with a faster progression. Juvenile forms are described as showing a paternal expansion bias. The chance of expansion of intermediate alleles is reported as 2%–10% in a paternal transmission.

Characteristics of Juvenile Huntington Disease

- Stiff, rigid muscles (without chorea)
- Seizures (25%–30% of patients with juvenile HD)
- Expanded repeat size 80–100 CAG
- 70%–90% Juvenile HD paternally inherited

Because of a high disease burden associated with HD, approved testing protocols have been developed in HD approved testing centers. Guidelines endorsed by the Huntington Disease Society of America and recommendations from ACMG/ASHG are described below.

- Pretest discussion with genetic counselor
    - Assessing patient's risk perception, expectations and support systems
    - Explaining implications of testing vs not testing
    - Medical management and reproductive options
- Performance of a neurological examination
- Psychologist/psychiatrist evaluation for depression, or other psychiatric illness
    - Confirmation of diagnosis of HD through molecular testing in an affected family member should be performed if possible before presymptomatic testing is offered
- Informed consent
    - Document informed consent for diagnostic and predictive testing
    - Consent must note whether the patient permits sample to be stored and/or used for research

Predictive testing on minors pose an ethical difficulty since HD is an adult-onset condition, and no treatment is currently available. Several societies or organizations have developed recommendations.

- General consensus: contraindicated
- Huntington Disease Society of America: Not recommended, but each center should develop their own policy
- International HD Association: Not recommended
- National Society of Genetic Counselors: Consider that many adults (80%) would choose not to have testing

Tests

PCR covers the CAG region of the gene. If results show an apparent homozygous in the normal range, an adjacent polymorphic CCG repeat region can be interrogated to ensure that two alleles are present. If CCG also apparently homozygous, Southern analysis can ensure that an expanded allele is not present.

Juveniles affected with HD will have greatly expanded alleles, and possibly not detected by PCR. Southern blot can confirm an expanded allele, although it will not be accurately sized.

Indications for Testing

HD testing is done to confirm diagnosis in symptomatic individuals, or predict disease. Presymptomatic testing is done for adults with a family history but without symptoms. Testing for nonsymptomatic juveniles is contraindicated because there is no treatment available.

Sample Type

DNA from whole blood (EDTA or ACD), saliva or cheek swabs, yield adequate DNA for PCR. If Southern analysis is needed, whole blood samples are recommended.

Table 7. CAG Repeats

Alleles	*CAG Repeats*	*Interpretation*
Normal	≤26	Unaffected
Mutable normal	27–35	Unaffected, offspring at risk for expansion
Reduced penetrance	36–39	At risk for developing symptoms
Full penetrance	≥40	Disease-causing

Results

Results are given as CAG repeat sizes for each allele.

Interpretation

The interpretations of the CAG repeats are given in Table 7.

Limitations

Although higher number of CAG repeats correspond with earlier age of onset, prediction of age of onset, rate of disease progression etc based on DNA results is not possible

Comments

A rare benign variant within primer regions may interfere with CAG amplification.

Hearing Loss

Hearing loss affects 1:500 (mild to moderate) to 1:1000 (severe to profound) infants. Statewide programs in the United States (US)

have newborn screening. Hearing loss can be due to environmental or inherited factors. Inherited hearing loss may be part of a syndrome or considered nonsyndromic.

- **Environment factors** (50% of deafness)
  - Ototoxic drugs
  - Acoustic trauma
  - Bacterial/viral infections
- **Inherited** (50% of deafness)
  - Syndromic hearing loss includes numerous diseases and genes and is responsible for approximately 30% of inherited hearing loss.
  - Nonsyndromic hearing loss (NSHL) includes over 70 genetic loci and is responsible for up to 70% of inherited hearing loss. Approximately 40 loci have been described for autosomal dominant (DFNA) forms that represent about 75% of NSHL, while autosomal recessive (DFNB) forms make up about 20%. X-linked and mitochondrial forms make up 1%–5% of inherited hearing loss.

Two genes coding for connexins (a membrane protein involved in the formation of channels) are responsible for the majority of NSHL. The most common gene responsible for NSHL is *GJB2* (connexin 26), which includes both recessive and dominant mutations. A common mutation, c.35delG, has been reported. Deletions of *GJB6* (connexin 30) have been reported, but homozygosity for a *GJB6* deletion hasn't been reported in individuals with hearing loss. Rather, it is the combination of a recessive *GJB2* mutation and a *GJB6* deletion that results in hearing loss.

Two mitochondrial mutations, m.1555A>G and m.7445A>G, have been described in NSHL. They show maternal inheritance patterns typical for mitochondrial mutations, with variable severity based on the amount of heteroplasmy present in the patient. The m.1555A>G mutation has been associated with use of aminoglycosides.

Tests

Sequencing for *GJB2* (exons and known 5′ mutations within regulatory regions) is available. If one recessive mutation is identified by sequencing, deletion analysis for *GJB6* should be performed. *GJB6* analysis typically detects two different gene deletions.

Indications for Testing

Indications for NSHL testing include carrier testing for family members. Figure 4 shows a testing algorithm for hearing loss.

Sample Type

DNA from whole blood (EDTA or ACD), blood spots, saliva, cheek swabs, amniotic fluid, or amniocytes yield adequate DNA for c.35delG or for family specific mutations. Whole blood samples or cultured amniocytes are recommended for full gene sequencing and deletion analysis.

Results

Results are given as negative for a mutation in *GJB2* and mitochondrial mutations, or positive (heterozygous, homozygous or compound heterozygous) for mutations. *GJB6* results will be positive or negative for a deletion.

Interpretation

Two recessive *GJB2* mutations, or one recessive *GJB2* mutation and one *GJB6* deletion, indicate the cause of hearing loss. One dominant *GJB2* mutation indicates the cause of hearing loss.

Limitations

Many other genes have been shown to contribute to hearing loss. A negative *GJB2* (and *GJB6*) does not rule out an inherited cause of hearing loss.

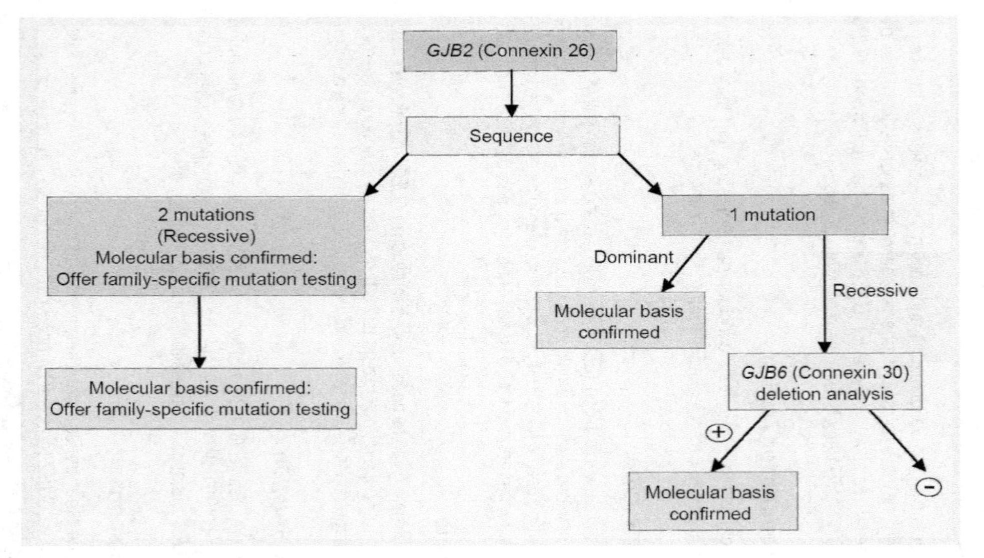

Figure 4. Algorithm for hearing loss.

Comments

Laboratories are beginning to offer large hearing loss gene panels, using next generation sequencing or array technologies.

Spinal Muscular Atrophy

Spinal muscular atrophy (SMA) is a lower motor neuron disease with onset predominantly in infancy or early childhood. It is characterized by symmetrical limb and trunk weakness, muscular atrophy and degeneration, and loss of the anterior proximal horn cells in the spinal cord. It is an autosomal recessive with an incidence of ~1:10,000 and a carrier frequency of ~1:100. Three types of SMA have been described, based on severity of disease. All three forms are linked to 5q13. The types and characteristics of SMA are listed in Table 8.

Table 8. Characteristics and Type of Spinal Muscular Atrophy

Type	*Characteristic*
1[a]	Most severe form with age of within first 6 months Severe generalized weakness at birth or shortly thereafter Never able to sit without support Death in first 2 y (most cases)
2	Cannot sit, stand, or walk unaided More prolonged survival
3	Able to stand and walk unaided Highly variable age of onset Usually show proximal muscle Weakness in early childhood
4	Adult onset Muscle weakness Normal lifespan

[a]Also known as Werdnig-Hoffman disease.

More than 95% of patients with SMA are homozygous for a microdeletion involving the *SMN* gene. The *SMN* gene has telomeric (SMN^T or *SMN1*) and centromeric (SMN^C or *SMN2*) copies apparently derived from an inverted duplication event; consequently, there are very slight differences (five bases) between *SMN1* and *SMN2* in exons 7 and 8. One difference in exon 7 (c.840C>T) accounts for decreased transcription from SMN2. Molecular analyses use this difference for diagnostic testing. Approximately 94% of patients affected with SMA show a deletion of this region of *SMN1*.

Genetic modifiers of the severity of disease include gene conversion of *SMN1* to *SMN2* and increased copies of *SMN2*. Deletions or mutations of *SMN2* alone do not produce SMA.

Carrier testing for deletions is also available, although the genetics are complicated by the presence of duplications of *SMN1*. A total of 82%–96% of normal individuals have one copy on each chromosome, while 4%–18% of individuals have two copies.

Tests

Molecular tests are based on detecting deletions in the *SMN1* gene. These assays use single base pair differences between *SMN1* and *SMN2* in exon 7 and/or exon 8. Several technologies are available, from restriction digest to Luminex or MLPA based assays. Sequencing the gene for affected individuals that do not have two deletions is available. Copy number determination of *SMN2* is available.

Indications for Testing

Molecular confirmation of affected individuals, newborn screening, carrier testing of family members of affected individuals, carrier screening. If affected individuals are not homozygous for deletions, full gene sequencing can be performed.

Sample Type

DNA from whole blood (EDTA or ACD), blood spots, saliva, cheek swabs, amniotic fluid or amniocytes yield adequate DNA for mutation panels or for family specific mutations. Whole blood samples are recommended for full gene sequencing.

Results

Results are given as no deletion detected, one (heterozygous) or two (homozygous) deletions detected. If sequencing is performed, results are given as the mutation detected and classified as to its significance.

Interpretation

Two copies of a deletion of *SMN1* are diagnostic for SMA. One copy of *SMN1* indicates at least a carrier status, and may be affected if a rare sequence change is present on the opposite chromosome. In very rare instances, an affected individual theoretically could carry two rare sequence variants on opposite chromosomes.

Limitations

SMA carriers of a deletion, yet having two copies of *SMN1* on the same chromosome will not be detected as carriers by current methods. Screening methods for affected individuals test for homozygous deletions only. Individuals with sequence mutations will not be identified as affected or as carriers.

Comments

This area of the genome has many repeated elements and pseudogenes, which can affect assay design.

Dystrophinopathies

Dystrophinopathies are a group of X-linked recessive disorders usually due to mutations in the dystrophin (*DMD*) gene. Severe mutations, those resulting in the absence of dystrophin protein, result in Duchenne muscular dystrophy (DMD). Onset of this disorder is typically less than 5 y of age. Milder mutations, such as in-frame deletions that produce some dystrophin protein, result in Becker muscular dystrophy (BMD). The dystrophin gene is large, with 79 exons. Large deletions involving the dystrophin gene can be detected by chromosome or genomic microarray analysis. For smaller deletions, exonic level deletion/duplication testing is needed. If no mutation is found, the entire gene can be examined for single base changes or small insertions/deletions by sequencing or mutation scanning.

Characteristics

- Muscle weakness
- Pseudohypertrophy of the calves
- Impaired heart and respiratory musculature
- Becker type is a milder expression of disease

Indications for Testing

The three common reasons for molecular testing for *DMD* mutations are:

- Confirming mutation in affected individuals
- Determine carrier status of females with a family history
- Fetal testing for family specific mutation

Sample Type

DNA from whole blood (EDTA or ACD), or saliva, cheek swabs, amniotic fluid, or amniocytes yield adequate DNA for family specific mutations. Whole blood samples or cultured amniocytes are recommended for full gene sequencing and deletion/duplication analysis,.

Results

The majority of mutations are deletions and duplications with the following mutation spectrum for DMD and BMD:

- **DMD**
    - Deletions: Approximately 65%
    - Duplications; Approximately 7%–10%
    - Point mutations/small insertions or deletions: Approximately 25%–30%
- **BMD**
    - Deletions: Approximately 85%
    - Duplications: Approximately 6%–10%
    - Point mutations/small insertions or deletions: Approximately 5%–10%

Interpretation

Large deletions or duplications are seen in both DMD and BMD. To predict the severity, the deletion or duplication needs to be evaluated for whether a frameshift (DMD) or an in-frame (BDM) deletion has occurred. Female carriers may have variable symptoms of muscular dystrophy.

Limitations

A karyotype or chromosomal microarray will detect large gene rearrangements of the X chromosome, but does not have exon-level resolution. Deletion/duplication analysis (either by genomic or exonic level) will not detect point mutations.

Comments

Large rearrangements may be detected coincidently when testing for development delay by chromosomes or genomic microarray. In these cases, the report may suggest confirming the deletion or duplication with higher resolution techniques.

Suggested Reading

Allingham-Hawkins DJ, Babul-Hirji R, Chitayat D, et al. Fragile X premutation is a significant risk factor for premature ovarian failure: The International Collaborative POF in Fragile X study – preliminary data. Am J Med Genet 1999;83:322–5.

American College of Medical Genetics. Technical standards and guidelines for Fragile X testing: a revision to the disease-specific supplements to the Standards and Guidelines for Clinical Genetics Laboratories of the American College of Medical Genetics. http://www.acmg.net/Pages/ACMG_Activities/stds-2002/fx.htm (Accessed May 8, 2013).

Amir RE, Van den Veyver IB, Wan M, et al. Rett syndrome is caused by mutations in X-linked MECP2, encoding methyl-CpG-binding protein 2. Nat Genet 1999;23:185–8.

Amos J, Feldman GL, Grody WW, et al. Guidelines: technical standards and guidelines for CFTR mutation testing. http://www.acmg.net/AM/Template.cfm?Section=Laboratory_Standards_and_Guidelines&Template=/CM/ContentDisplay.cfm&ContentID=6328 (Accessed May 8, 2013).

Amos Wilson J, Pratt VM, Phansalkar A, et al. Consensus characterization of 16 FMR1 reference materials: a consortium study. J Mol Diagn 2008;10:2–12.

Ballana E, Ventayol M, Rabionet R, Gasparini P, Estivill X. Connexins and deafness. http://davinci.crg.es/deafness/ (Accessed May 8, 2013).

Beggs AH, Hoffman EP, Snyder JR et al. Exploring the molecular basis for variability among patients with Becker muscular dystrophy: dystrophin gene and protein studies. Am J Hum Genet 1991;49:54–67.

Castellani C, Cuppens H, Macek M Jr, et al. Consensus on the use and interpretation of cystic fibrosis mutation analysis in clinical practice. J Cyst Fibros 2008;7:179–96.

Chillón M, Casals T, Mercier B, et al. Mutations in the cystic fibrosis gene in patients with congenital absence of the vas deferens. N Engl J Med 1995;332:1475–80.

Clark BE, Thein SL. Molecular diagnosis of hemoglobin disorders. Clin Lab Haematol 2004;26:159–76.

Cystic fibrosis mutation database. http://www.genet.sickkids.on.ca/app (Accessed February 12, 2013).

Derry S, Wood WG, Pippard M, et al. Hematologic and biosynthetic studies in homozygous hemoglobin constant spring. J Clin Invest 1984;73:1673–82.

Galanello R, Cao A. Alpha-thalassemia. Genet Med 2011;1383–8.

Gitlin JM, Fischbeck K, Crawford TO, et al. Carrier testing for spinal muscular atrophy. Genet Med 2010;12:621–2.

Groman JD, Hefferon TW, Casals T, et al. Variation in a repeat sequence determines whether a common variant of the cystic fibrosis transmembrane conductance regulator gene is pathogenic or benign. Am J Hum Genet 2004;74:176–9.

Hardwick SA, Reuter K, Williamson SL, et al. Delineation of large deletions of the MECP2 gene in Rett syndrome patients, including a familial case with a male proband. Eur J Hum Genet 2007;15:1218–29.

Hegde MR, Chin EL, Mulle JG, et al. Microarray-based mutation detection in the dystrophin gene. Hum Mutat 2008;29:1091–9.

Hickey SE, Curry CJ, Toriello HV. ACMG practice guideline: lack of evidence for MTHFR polymorphism testing. Genet Med 2013;15:153–6.

Higuchi M, Kazazian HH Jr, Kasch L, et al. Molecular characterization of severe hemophilia A suggests that about half the mutations are not within the coding regions and splice junctions of the factor VIII gene. Proc Natl Acad Sci USA 1991;88:7405–9.

Huntington's Disease Society of America. Guidelines for genetic testing for Huntington's disease. http://www.hdfoundation.org/html/hdsatest.php (Accessed May 8, 2013).

Kembell-Cook G. Haemophilia A mutation database. http://hadb.org.uk (Accessed May 8, 2013).

King C, Barton DE. Best practice guidelines for the molecular genetic diagnosis of type I (HFE-related) hereditary haemochromatosis. BMC Med Genet 2006;7:81.

Kleefstra T, Yntema HG, Nillesen WM, et al. MECP2 analysis in mentally retarded patients: implications for routine DNA diagnostics. Eur J Hum Genet 2004;12:24–8.

Kluijtmans LA, van den Heuvel LP, Boers GH, et al. Molecular genetic analysis in mild hyperhomocysteinemia: a common mutation in the methylenetetrahydrofolate reductase gene is a genetic risk factor for cardiovascular disease. Am J Hum Genet 1996;58:35–41.

Langbehn DR, Brinkman RR, Falush D, et al. A new model for prediction of the age of onset and penetrance for Huntington's disease based on CAG length. Clin Genet 2004;65:267–77.

Liu M, Murphy ME, Thompson AR. A domain mutations in 65 haemophilia A families and molecular modeling of dysfunctional factor VIII proteins. Br J Haemotol 1998;103:1051–60.

Liu ML, Nakaya S, Thompson AR. Non-inversion factor VIII mutations in 80 hemophilia A families including 24 with alloimmune responses. Thromb Haemost 2002;87:273–6.

Lyon E, Laver T, Ping Y, et al. A simple, high throughput assay for Fragile X expanded alleles using triple repeat primed PCR and capillary electrophoresis. J Mol Diagn 2010;12:505–11.

Martinelli I, Bucciarelli P, Margaglione M, et al. The risk of venous thrombosis in family members with mutations in the genes of Factor V or prothrombin or both. Br J Haematol 2000;111:1223–9.

Miltenberger-Miltenyi G, Laccone F. Mutations and polymorphisms in the human methyl CpG-binding protein MECP2. Hum Mutat 2003;22:107–15.

Murray A. Premature ovarian failure and the FMR1 gene. Semin Reprod Med 2000;18:59–66.

Nicolaes GA, Dahlback B. Activated protein C resistance (FV(Leiden)) and thrombosis: Factor V mutations causing hypercoagulable states. Hematol Oncol Clin North Am 2003;17:37–61, vi.

Panigrahi I, Kesari A, Phadke SR, et al. Clinical and molecular diagnosis of spinal muscular atrophy. Neurol India 2002;50:117–22.

Peixeiro I, Silva AL, Romão L. Control of human beta-globin mRNA stability and its impact on beta-thalassemia phenotype. Haematologica 2011;96:905–13.

Pollak A, Skórka A, Mueller-Malesińska M, et al. M34T and V37I mutations in GJB2 associated hearing impairment: Evidence for pathogenicity and reduced penetrance. Am J Med Genet A 2007;143A:2534–43.

Poort SR, Rosendaal FR, Reitsma PH, et al. A common genetic variation in the 3'- untranslated region of the prothrombin gene is associated with elevated plasma prothrombin levels and an increase in venous thrombosis. Blood 1996;88:3698–703.

Potter NT, Spector EB, Prior TW. Technical standards and guidelines for Huntington disease testing. Genet Med 2004;6:61–5.

Prior TW, Russman BS; University of Washington. Spinal muscular atrophy. http://www.ncbi.nlm.nih.gov/books/NBK1352 (Accessed May 8, 2013).

Ravn K, Nielsen JB, Skjeldal OH, et al. Large genomic rearrangements in MECP2. Hum Mutat 2005;25:324.

Rossi E, Jeffrey G. Clinical penetrance of C282Y homozygous HFE haemochromatosis. Clin Biochem Rev 2004;25:183–90.

Rost S, Löffler S, Pavlova A, et al. Detection of large duplications within the factor VIII gene by MLPA. J Throm Haemost 2008;6:1996–8.

Snoeckx RL, Huygen PL, Feldmann D, et al. GJB2 mutations and degree of hearing loss: a multicenter study. Am J Hum Genet 2005;77:945–957.

Spector EB, Grody WW, Carla J. Matteson CJ, et al. Technical standards and guidelines: venous thromboembolism (Factor V Leiden and prothrombin 20210G>A testing): a disease-specific supplement to the standards and guidelines for clinical genetics laboratories. Genet Med 2005;7:444–53.

Strom CM, Crossley B, Redman JB, et al. Molecular testing for Fragile X Syndrome: lessons learned from 119,232 tests performed in a clinical laboratory. Genet Med 2007;9:46-51

Tang HY, Fang P, Ward PA, et al. DNA sequence analysis of GJB2, encoding connexin 26: Observations from a population of hearing impaired cases and variable carrier rates, complex genotypes, and ethnic stratification of alleles among controls. Am J Med Genet A 2006;140:2401–15.

Thomas MW, McInnes RR, Willard HF. The Hemoglobinopathies: Models of Molecular Disease. In: Nussbaum RM, McInnes RR, Willard HF, eds. Genetics in Medicine, 5th ed. Philadelphia: WB Saunders Co., 1991:247–70.

Van Esch H, Bauters M, Ignatius J, et al. Duplication of the MECP2 region is a frequent cause of severe mental retardation and progressive neurological symptoms in males. Am J Hum Genet 2005;77:442–53.

Weiss FU, Simon P, Bogdanova N, et al. Complete cystic fibrosis transmembrane conductance regulator gene sequencing in patients with idiopathic chronic pancreatitis and controls. Gut 2005;54:1456–60.

Wu BL, Lindeman N, Lip V, et al. Effectiveness of sequencing connexin 26 (GJB2) in cases of familial or sporadic childhood deafness referred for molecular diagnostic testing. Genet Med 2002;4:279–88.

Molecular Oncology

Oncology testing is one of the fastest growing areas in the practice of molecular diagnostics. Where once molecular testing was extremely rare, it is now becoming the standard of care in many forms of neoplasia. Molecular oncology testing can be divided into testing relating to:

- **Diagnostic testing:** Testing that is performed in order to confirm a suspected cancer diagnosis or to differentiate between potential cancer types in a differential.
- **Prognostic testing:** Testing that is performed to determine the status of molecular markers of prognosis in patients with a given disease.
- **Therapeutic target testing:** Testing is performed to determine which therapeutic method should be implemented in a given patient.

In many situations these categories of testing overlap as some testing provides a large amount of clinical information. In this section, testing for several of the most commonly observed neoplastic disorders will be discussed. Where possible, diagnostic, prognostic, and therapeutic targets testing relating to each order will be addressed. Sample requirements, results and interpretation will also be discussed.

Myeloproliferative Disorders

Chronic Myelogenous Leukemia

Chronic myelogenous (or myeloid) leukemia (CML) is a myeloproliferative disorder accounting for approximately 15%–20% of all cases of leukemia. At the cellular level, CML is marked by the clonal proliferation of myeloid cells, most notably the granulocytes, in the bone marrow. At the molecular level, CML is defined by a chromosomal translocation involving the long arms of chromosome 9 (9q34.1) and 22 (22q11) that is known as the Philadelphia chromosome. This translocation is almost invariably observed in patients with CML and results in the formation of a fusion gene between the *BCR* gene (located on chromosome 22) and the *ABL1* gene (located on chromosome 9). The product of this fusion gene is a constitutively activated protein kinase known as BCR/ABL.

Characteristics

- Unexplained weight loss
- Loss of appetite
- Lethargy
- Night sweats
- Easy bleeding
- Frequent infections
- Splenomegaly

Tests

Several molecular tests are commonly performed to confirm a diagnosis of CML. These are typically done after an abnormal complete blood count (CBC) and/or the presence of splenomegaly have increased the clinical suspicion for CML. In addition, several tests exist in order to aid in monitoring treatment efficacy in patients with this disorder. Examples of these tests are described below.

- ***BCR-ABL1* fluorescence *in situ* hybridization (FISH)**
    - A cytogenetic method that utilizes fluorescently labeled probes specific for both the *BCR* and *ABL1* genes to quickly identify the presence of the *BCR-ABL1* fusion gene.
    - This test is commonly used to rapidly diagnose CML and can also be used to monitor disease after treatment has been initiated. However, this method is not suited for the detection of minimal residual disease (MRD).

Sample Type

Bone marrow (sodium heparin) is the preferred specimen.

Limitations

A false-negative result is possible in samples with low levels of cancer cells.

- **Qualitative reverse-transcriptase polymerase chain reaction (RT-PCR) for *BCR-ABL1***
  - Several different breakpoints have been described in the *BCR* gene that result in variability of the protein product produced by the fusion transcript.
  - The most commonly occurring (major) breakpoint results in a protein product that is 210 kDa. Less commonly, the breakpoint results in a 190-kDa product. The 190-kDa product is more commonly observed in patients with acute lymphoblastic leukemia (ALL).
  - Qualitative RT-PCR is one method that is routinely used to differentiate between the two most common products observed in patients with CML. Determination of the product size is critical for future monitoring of a patient's disease status by molecular methods.

Sample Type

Bone marrow (ethylenediaminetetraacetic acid [EDTA]) and peripheral blood (EDTA). However, these samples must be obtained by the laboratory within 48 h due to the delicate nature of the RNA used in this analysis method.

Limitations

A very rare *BCR* breakpoint results in a 230-kDa product that is not routinely tested for in most clinical molecular laboratories and may not be detected.

- **Quantitative real-time PCR (qPCR) for *BCR-ABL1***
  - qPCR is the most commonly used method to monitor treatment efficacy in patients with CML.
  - Typically, this test is performed at the time of diagnosis/initiation of therapy. This allows for the determination of baseline tumor burden. Subsequently, this test is performed after the patient has begun a treatment regimen. Decreases in the *BCR-ABL1* copy number indicate that treatment is effective in reducing tumor burden while increases in copy number indicate that treatment should be modified.
  - qPCR is a highly sensitive method capable of detecting as few as 10 copies of the *BCR-ABL1* transcript in a given sample and can be performed to specifically detect transcripts producing both the 190-kDa and 210-kDa products. Therefore, qPCR is the ideal method to screen for MRD.

Sample Type

Both bone marrow (EDTA) and peripheral blood (EDTA) are acceptable. However, these samples must be obtained by the laboratory within 48 h due to the delicate nature of the RNA used in this analysis method.

Limitations

A very rare *BCR* breakpoint results in a 230-kDa product that is not routinely tested for in most clinical molecular laboratories and may not be detected.

- ***BCR-ABL1* kinase domain mutation testing**
  - The most common treatment modality in patients with CML is the use of tyrosine kinase inhibitors (TKIs).
  - Although TKIs are extremely effective in decreasing tumor burden initially, they lose their effectiveness over time. It has been recently determined that a large degree of drug resistance is due to point mutations located in the *ABL1* kinase domain. Specific

point mutations have been identified that are associated with resistance to various TKIs (e.g., the p.Thr315Ile mutation). Therefore, sequence analysis of this region is commonly performed to determine if mutations in this region are the cause of the loss in treatment efficacy and to redirect the treatment method.

Sample Type

Bone marrow (EDTA) and peripheral blood (EDTA).

Limitations

Other factors that contribute to TKI resistance/response are not detected by this assay.

Acute Myeloid Leukemia

Acute myeloid (or myelogenous) leukemia (AML) is a myeloproliferative disease characterized by the uncontrolled proliferation of hematopoietic progenitor cells. AML is an exceptionally heterogeneous disorder with many subtypes that are delineated by morphology, immunophenotyping, histochemical analysis, and the presence of genetic abnormalities. AML is a devastating disorder with an overall long-term survival of approximately 30%.

Characteristics

- Lethargy
- Bone pain
- Easy bruising
- Frequent infection
- Abnormal bleeding
- Shortness of breath
- Anemia
- Thrombocytopenia

Tests

Molecular diagnostic testing in patients with AML has become commonplace over the last decade as several molecular markers have been identified that provide prognostic information as well as a marker for tracking minimal residual disease in a patient. Although some molecular testing is available to help classify the AML subtype, the majority of molecular testing performed on patients with AML is used to help in establishing prognosis and treatment modalities. Examples of these types tests are described below. It should be noted that most the tests described below are typically performed after a diagnosis of AML has been established.

- ***PML/RARA* fusion gene testing**
  - Acute promyelocytic leukemia (APL) is a subtype of AML defined by a translocation between the long arms of chromosomes 15 (15q24) and 17 (17q21).
  - The t(15;17) observed in APL results in the formation of a fusion oncogene between the *RARA* gene on chromosome 17 and the *PML* gene on chromosome 15.
  - The *PML/RARA* fusion gene produces a protein with altered function that ultimately results in immortalized cells incapable of terminal differentiation.
  - Molecular testing for the *PML/RARA* fusion gene is performed by RT-PCR followed by any number of detection methods. Many laboratories have begun testing for this fusion gene by quantitative RT-PCR methods to monitor patient response to therapy over time.
  - Although morphologic evaluation is essential in determining which patients should be tested for the *PML/RARA* fusion, molecular detection of this mutation is essential for definitive diagnosis of APL.
  - The detection of the *PML/RARA* fusion gene has significant implications on prognosis and treatment. Patients with this genetic alteration respond remarkably well to retinoic acid

based therapy. In fact, the use of retinoic acid based therapies is able to cure as many as 90% of patients with APL.

Sample Type

Both bone marrow (EDTA) and peripheral blood (EDTA) are acceptable. However, these samples must be obtained by the laboratory within 48 h due to the delicate nature of the RNA used in this analysis method.

Limitations

Subtypes of AML other than APL will not be detected by this assay.

- ***FLT3* internal tandem duplication (ITD) testing**
  - Mutations in the FMS-like tyrosine kinase-3 (*FLT3*) gene are observed in 25% of adult patients with AML (the percentage is higher in patients with a normal karyotype).
  - The most common mutation in the *FLT3* gene is an ITD in the juxtamembrane domain of the gene and results in the constitutive activation of the FLT3 tyrosine kinase.
  - *FLT3*-ITD is an exact duplication of a varying number of base pairs of the gene. It is important to note that these duplications are always in frame. Out of frame mutations will not produce a constitutively activated FLT3 protein.
  - Several methods exist for the detection of *FLT3*-ITD mutations but the most common method is PCR amplification of the juxtamembrane region of the gene followed by capillary electrophoresis. A patient that is positive for an *FLT*-ITD mutation will exhibit (in the vast majority of cases) both a wild-type allele and one that is larger in size but in frame.
  - The presence of an *FLT3*-ITD mutation confers a negative prognosis and may indicate that the patient is unlikely to respond favorably to traditional chemotherapy.

- Clinical trials investigating the use of tyrosine kinase inhibitors that specifically target the constitutively activated FLT3 kinase are currently underway.

Sample Type

The preferred initial specimen is a diagnostic bone marrow (EDTA). After the initial testing of the diagnostic specimen, a peripheral blood (EDTA) or bone marrow (EDTA) may be submitted.

Limitations

A low number of malignant cells (below the limit of assay detection) in a sample can result in a false-negative result.

- ***FLT3* kinase domain mutation testing**
  - Point mutations in the *FLT3* tyrosine kinase domain (TKD) occur in 5%–7% of adult patients with AML.
  - The majority of these mutations are identified in the activation loop of the TKD at codons 835 and 836.
  - Several methods are available to identify *FLT3*-TKD mutations. Some laboratories specifically target codons 835 and 836 by allele-specific PCR (AS-PCR) while others perform PCR followed by Sanger sequencing of the entire TKD to increase the sensitivity of the test.
  - The prognostic significance of *FLT3*-TKD mutations is contentious and is the topic of many ongoing studies.

Sample Type

The preferred initial specimen is a diagnostic bone marrow (EDTA). After the initial testing of the diagnostic specimen, a peripheral blood (EDTA) or bone marrow (EDTA) may be submitted.

Limitations

A low number of malignant cells (below the limit of assay detection) in a sample can result in a false-negative result. In addition, targeted testing does not exclude the possibility of a disease-causing mutation not detected by the assay.

- ***NPM1* mutation testing**
  - Mutations in the *NPM1* gene are detected in 25%–35% of adult patients with AML.
  - The most common mutation detected in the *NPM1* gene is a 4 bp insertion in exon 12 of the gene that interrupts the nuclear localization signal resulting in the aberrant cytoplasmic localization of the protein.
  - Molecular testing for the 4 bp insertion mutation is typically performed by PCR specific for *NPM1* exon 12 followed by capillary electrophoresis. The mutated product is easily detectable as it is 4 bp longer than that of the wild-type allele.
  - The presence of an *NPM1* mutation confers a favorable prognosis (in patients without concomitant *FLT3*-ITD gene mutations). In addition, studies have indicated that *NPM1*$^{+}$ (*FLT3*-ITD^{-}) has a positive response to traditional induction chemotherapy.

Sample Type

The preferred initial specimen is a diagnostic bone marrow (EDTA). After the initial testing of the diagnostic specimen, a peripheral blood (EDTA) or bone marrow (EDTA) may be submitted.

Limitations

A low number of malignant cells (below the limit of assay detection) in a sample can result in a false-negative result.

- ***CEBPA* mutation testing**
  - Mutations in the *CEBPA* gene are observed in 5%–15% of patients with AML.
  - Two types of mutations occur in the *CEBPA* gene:
    - N-terminal nonsense mutations that result in a dominant negative CEBPA isoform
    - In-frame insertions/deletions in the C-terminal region of the gene that result in a CEBPA protein with reduced binding activity
  - As mutations are located throughout the *CEBPA* gene, molecular testing typically involves the use of PCR and the subsequent Sanger sequencing of the gene.
  - The presence of two (biallelic) mutations in the *CEBPA* gene confers a favorable prognosis (in the absence of a concomitant *FLT3*-ITD mutation). In addition, the presence of biallelic *CEBPA* mutations (in the absence of a *FLT3*-ITD mutation) indicates that the patient is likely to respond favorably to traditional induction chemotherapy.

Sample Type

The preferred initial specimen is a diagnostic bone marrow (EDTA). After the initial testing of the diagnostic specimen, a peripheral blood (EDTA) or bone marrow (EDTA) may be submitted.

Limitations

A low number of malignant cells (below the limit of assay detection) in a sample can result in a false-negative result.

Polycythemia Vera

Polycythemia vera (PV) is a myeloproliferative disorder that results in an abnormal increase in production of red blood cells (and some other blood cell types) by the bone marrow.

Table 9. Clinical Significance of Molecular Alterations in Acute Myeloid Leukemia

Gene Mutation Detected	*Prognostic Significance*
PML/RARA fusion gene	Associated with a favorable clinical outcome; 90% of patients with this mutation are cured by retinoic acid therapy
FLT3-ITD+	Associated with poor clinical outcome and poor response to traditional chemotherapy
FLT3-TKD+	Association with clinical outcome uncertain
NPM1 4bp insertion	In the absence of a concomitant *FLT3*-ITD mutation the presence of an *NPM1* mutation confers a favorable prognosis and indicates a favorable response to traditional chemotherapy
CEBPA	In the absence of a concomitant *FLT3*-ITD mutation the presence of biallelic *CEBPA* mutations confers a favorable prognosis and indicates a favorable response to traditional chemotherapy

Characteristics

- Fatigue
- Dizziness
- Shortness of breath
- Recurrent headache
- Enlarged spleen
- Abnormal and excessive bleeding
- Shortness of breath

Essential Thrombocythemia

Essential thrombocythemia (ET) is a myeloproliferative disorder that results from the abnormally high production of platelets by the bone marrow.

Characteristics

- Fatigue
- Dizziness
- Chest pain
- Recurrent headache
- Enlarged spleen

Myelofibrosis

Myelofibrosis (MF) is a myeloproliferative disorder characterized by the replacement of bone marrow with fibrotic tissue. MF can be either a primary disorder or the result of the progression of PV or ET.

Characteristics

- Anemia
- Lethargy
- Easy bleeding/bruising
- Enlarged spleen
- Bone pain
- Shortness of breath

Tests

Molecular testing to aid in the diagnosis of PV, ET, and MF has become commonplace over the last few years. This is largely due to the discovery that mutations in the *JAK2* gene contribute significantly to the pathogenesis of each of these disorders. Variations in molecular testing of the *JAK2* gene in each of these disorders are described in detail below. In addition, other markers observed in some of these disorders are also discussed.

- ***JAK2* V617F mutation testing**
  - Mutations in the *JAK2* gene are identified in 99% of patients with PV and 60% of patients with ET and MF.
  - The *JAK2* gene produces a protein with an important role in cell signaling. Clinically relevant mutations in this gene result in a gain-of-function, ultimately causing clonal expansion of certain cell types.
  - 96% of the *JAK2* mutations observed in PV and the vast majority of those observed in ET patients involve an amino acid alteration of valine to phenylalanine at codon 617 (V617F).
  - Molecular testing for the *JAK2* V617F mutation is typically performed by allele specific PCR (both qualitative and quantitative) followed by a number of detection methodologies.
  - The presence of the *JAK2* V617F mutation in a patient with symptoms of PV confirms a diagnosis.
  - The presence of the *JAK2* V617F mutation in ET confers an increased risk for thrombotic events and a decreased risk that ET will progress to myelofibrosis.
  - The presence of the *JAK2* V617F mutation in MF patients may help direct therapy as JAK2 kinase inhibitors (e.g., ruxolitinib) have been approved by the US Food and Drug Administration (FDA) for the treatment of *JAK2* mutation–related MF. In addition, clinical trials for the use of JAK2 kinase inhibitors in PV and ET are ongoing.

Sample Type

Peripheral blood (EDTA) or bone marrow (EDTA).

Limitations

Molecular detection of the *JAK2* V617F mutation is not diagnostic for any single myeloproliferative disorder and must be taken in context with histologic findings. In addition, a negative result by targeted *JAK2* V617F mutation testing does not exclude the possibility of a *JAK2* mutation as other disease-causing *JAK2* mutations may be present that will not be detected.

- ***JAK2* exon 12 mutation testing**
  - 3% of patients with PV have *JAK2* mutations located in exon 12 of the gene.
  - *JAK2* exon 12 mutations are rare in patients with ET and MF.
  - *JAK2* exon 12 mutations are typically small deletions and duplications.
  - Molecular testing for *JAK2* exon 12 mutations is typically performed by PCR for *JAK2* exon 12 followed by Sanger sequencing.
  - Because the V617F mutation testing detects the vast majority of *JAK2* mutations, exon 12 mutation testing is typically only recommended for patients who have tested negative for V617F.

Sample Type

Peripheral blood (EDTA) or bone marrow (EDTA).

Limitations

Disease-causing *JAK2* mutations located outside of exon 12 will not be detected by this assay.

- ***MPL* W515L/K mutation testing**
  - Mutations in the *MPL* gene occur in approximately 5% of patients with ET and 5%–10% of patients with MF.
  - *MPL* mutations are rare in PV.

- The most common mutation observed is a missense mutation at codon 515 of the gene. Mutations at this codon result in a change from tryptophan to either lysine or leucine.
- Molecular testing for *MPL* mutations in patients with MF is typically performed using allele specific PCR followed by capillary electrophoresis (or any number of detection methodologies). An alternative methodology for *MPL* mutation testing that is becoming commonplace is pyrosequencing. Pyrosequencing is a sequencing-by-synthesis method that allows for the detection of a mutation in as few as 5% of the cells in a sample (very useful in cases where the mutation of interest is somatic).
- Detected of an *MPL* mutation may support the diagnosis of MF or ET but a negative result does not exclude it.

Sample Type

Peripheral blood (EDTA) or bone marrow (EDTA).

Limitations

A negative result by targeted testing does not exclude the possible presence of another disease-causing *MPL* gene mutation (or causative mutations in other genes).

Solid Tumors

Breast Cancer

Breast cancer is the most common form of cancer observed in women. In fact, one in every eight women in the United States (US) will develop breast cancer in her lifetime. However, due to increased awareness and better screening methods breast cancer mortality is steadily decreasing.

Characteristics

- Breast lump/mass
- Breast asymmetry
- Breast pain
- Unexplained weight loss
- Unexplained nipple fluid discharge
- Arm swelling
- Bone pain

Tests

Traditionally, diagnostic testing for breast cancer has involved direct microscopic and immunohistochemical evaluation of tumor tissue. While this methodology is still the gold standard for determining a breast cancer diagnosis and classification, molecular methods are routinely being utilized to help determine treatment modalities, define risk and aid in the counseling of at-risk family members. In addition, presymptomatic molecular testing for some of hereditary breast cancer related gene mutations has become commonplace. Examples of each these types of molecular testing are discussed in detail below.

- ***ERBB2* (*HER2*) gene amplification FISH testing**
  - The *ERBB2* gene produces the human epidermal growth factor receptor 2 (HER2) protein, a receptor tyrosine kinase.
  - *ERBB2* gene amplification results in HER2 overexpression in 20%–30% of breast cancers.
  - Overexpression of HER2 results in enhanced cell signaling in pathways involved in cell proliferation, motility, and survival.
  - HER2 overexpression is associated with poor prognosis.
  - Molecular detection of HER2 is typically performed using FISH on breast tumor tissue.

- Detection of HER2 overexpression can help direct treatment as several HER2 targeting therapies have been developed (e.g., trastuzumab, lapatinib).

Sample Type

Formalin-fixed paraffin-embedded tumor tissue.

Limitations

Other factors that can contribute to drug response and clinical outcome are not detected by this assay.

- **Cytochrome P450 2D6 (*CYP2D6*) genotyping**
  - Cytochrome P450s are a group enzymes involved in drug metabolism.
  - The *CYP2D6* gene produces an enzyme that is critical in the metabolism of the chemotherapeutic drug tamoxifen and its active metabolite, endoxifen (as well as many other compounds).
  - Genetic variations in the *CYP2D6* gene are responsible for differences in the metabolism of many drugs (including tamoxifen and endoxifen).
  - Four distinct phenotypes have been observed based on *CYP2D6* genotypes:
    1. Normal (or extensive) metabolizers: Those with normal enzymatic activity.
    2. Intermediate metabolizers: Those with two alleles resulting in reduced enzymatic activity.
    3. Ultrarapid metabolizers: Those with more than two functional alleles resulting in increased enzymatic activity.
    4. Poor metabolizers: Those with two nonfunctional alleles and therefore markedly reduced enzymatic activity.

- Numerous studies have indicated that individuals classified as poor metabolizers do not respond well to tamoxifen therapy and thus helps direct therapy.
- Molecular genotyping of the *CYP2D6* gene is complex as more than 80 different alleles have been described.
- Several high quality, commercial kits are available that specifically target the most commonly observed *CYP2D6* alleles. These kits utilize a wide variety of methods that allow each laboratory to select the kit that will best fit into its normal workflow.

Sample Type

Peripheral blood (EDTA) is the preferred sample type.

Limitations

Not all *CYP2D6* alleles are detected by most commercially available kits. Therefore, some rare genotypes will not be detected.

- ***BRCA1* and *BRCA2* full gene testing**
  - Hereditary breast cancer refers to the form of breast cancer that is linked to inherited mutations. Overall, hereditary breast cancer represents between 5%–10% of breast cancer cases.
  - The two most frequently mutated genes in individuals with hereditary breast cancer are *BRCA1* and *BRCA2*.
  - Together, *BRCA1* and *BRCA2* account for approximately 10%–15% of all hereditary breast cancer cases.
  - A mutation in either of these two genes dramatically increases the risk of breast cancer over that of the general population and suggests that routine screening is recommended.
  - Presymptomatic testing for *BRCA1*/*BRCA2* mutations is recommended in individuals with a strong family history indicative of a hereditary breast cancer syndrome.

- *BRCA1/BRCA2* mutations are located throughout the entire gene. Therefore, molecular testing for *BRCA1/BRCA2* gene mutations involves Sanger sequencing of the entire coding region and intron/exon boundaries of the gene followed by capillary electrophoresis.
- Individuals not found to have a pathogenic mutation by Sanger sequencing are often tested for large deletions or duplications in the *BRCA1* and *BRCA2* genes by multiplex ligation-dependent probe amplification (MLPA).
- It is important to note that while identification of a *BRCA1* or *BCRA2* mutation increases the risk of breast cancer, it is not diagnostic and does not indicate that an individual will develop cancer.

Sample Type

Peripheral blood (EDTA) is the preferred sample type.

Limitations

A negative result does not exclude an increased cancer risk due to mutations in other genes that may contribute to the development of breast cancer.

- ***BRCA1* and *BRCA2* targeted mutation testing**
  - Persons of Ashkenazi Jewish (AJ) heritage are at risk of carrying a founder mutation in either the *BRCA1* or *BRCA2* genes.
  - The carrier frequency for *BRCA1/BRCA2* gene mutations among the AJ has been reported to be as high as 1:40.
  - Two AJ founder mutations, c.68_69delAG and c.5266dupC, have been described in the *BRCA1* gene and a single AJ founder mutation, c.5946delT, has been described in the *BRCA2* gene.
  - Molecular testing for the common AJ mutations in *BRCA1* and *BRCA2* is typically performed by targeted Sanger sequencing followed by capillary electrophoresis. However, many additional methods are capable of detecting these mutations.

Sample Type

Peripheral blood (EDTA) is the preferred sample type.

Limitations

The lack of a detection of mutation in *BRCA1* or *BRCA2* by targeted mutation testing does not eliminate the possibility that the patient is a carrier of a *BRCA1* or *BRCA2* gene mutation not detected by targeted mutation testing.

Colorectal Cancer

Colorectal (or colon) cancer is the third most common cancer diagnosed in the US. It is also the third most common cause of cancer death. However, mortality from colorectal cancer continues to decrease as screening tests are now considered the standard of care. Colorectal cancer is a heterogeneous disorder with many different causes (both genetic and environmental). This section will focus on inherited forms of colorectal cancer and the role that molecular diagnostics plays in the diagnosis of each disorder. In addition, molecular markers that aid in determining treatment modalities will also be discussed.

Characteristics

- Bloody stool/rectal bleeding
- Marked constipation
- Unexplained weight loss
- Abdominal pain
- Loss of appetite

Hereditary Nonpolyposis Colorectal Cancer

Hereditary nonpolyposis colorectal cancer (HNPCC), which is also known as Lynch Syndrome, is a form of inherited colon cancer caused by mutations in mismatch repair (MMR) genes. HNPCC is inherited in an autosomal dominant manner and is the most common hereditary cause of colorectal cancer (and other gastrointestinal [GI] cancers). In fact, several guidelines (e.g., the Amsterdam Criteria and the Revised Bethesda; Tables 10 and 11) have been developed to aid in identifying individuals likely to be affected with HNPCC and recommend these patients for testing. However, many believe that these guidelines are too broad and that HNPCC testing should be performed on all newly diagnosed cases of colorectal cancer.

Tests

Molecular testing in patients with a suspected diagnosis of HNPCC involves a tiered system (see Figure 5) whereby immunohistochemistry (IHC) is performed to determine if a specific MMR protein is absent from a tumor specimen. An abnormal IHC test result can indicate the type of molecular testing that is necessary and therefore should always be performed prior to molecular analysis (if tumor tissue is available) in a symptomatic individual. Molecular testing in HNPCC is considered the gold standard in confirming a diagnosis and is described in detail below.

Table 10. Revised Amsterdam Criteria for Hereditary Nonpolyposis Colorectal Cancer

≥ 3 relatives with hereditary nonpolyposis colorectal cancer:
One should be a first-degree relative of the other two
≥ 2 successive affected generations
≥ 1 individual diagnosed before age 50 y
Diagnosis of familiar adenomatous polyposis excluded
Tumors verified by histopathology

Table 11. Revised Bethesda Guidelines for Microsatellite Instability Testing
Individuals should undergo MSI testing if:
Colorectal cancer diagnosed < age 50 y
Multiple synchronous or metachronous HNPCC-related tumors present at any age
Colorectal cancer with MSI High histology present < age 60 y
Colorectal cancer diagnosed in ≥ 1 first-degree relatives with HNPCC-related tumor, with ≥ 1 of these cancers occurring < age 50 y
Colorectal cancer diagnosed in ≥ 2 first- or second-degree relatives at any age

HNPCC, hereditary nonpolyposis colorectal cancer; MSI, microsatellite instability.

- **Microsatellite instability (MSI) testing**
  - HNPCC is caused by mutations in MMR genes. These genes are critical in identifying DNA replication errors and repairing them. When mutations occur in MMR genes this "proofreading" function is lost.
  - Microsatellites, DNA repeats ranging in size from 1 to 6 base pairs, are common sites of DNA replication errors (increases or decreases of repeat number). MMR proteins repair these variations in a normal cell but are incapable of doing so in patients with HNPCC resulting in MSI.
  - MSI is a hallmark feature of HNPCC tumors but is also observed in some sporadic colon cancers.
  - MSI molecular testing is often the first molecular test performed in individuals with a suspected diagnosis of HNPCC that have normal IHC results.
  - MSI molecular testing is typically performed by PCR amplification of a series of five microsatellites (also known as short

tandem repeats [STRs]) in both a normal sample and a tumor specimen. The results of the normal sample are used for comparison with the tumor sample. Contractions or expansions in the microsatellites indicate a positive result while an exact match between normal and tumor indicate a negative.

- Results of MSI molecular testing are defined by the number of microsatellites showing instability.
  - *MSI High* designates instability observed in 2 or more of the microsatellites tested.
  - *MSI Indeterminate* designates instability observed in a single microsatellite.
  - *MSI Stable* indicates that no MSI was observed in the tumor and that a diagnosis of HNPCC is unlikely.
- Individuals with a MSI High or MSI Intermediate result should be considered for testing of MMR genes.
- Commercial testing kits are available for MSI PCR.

Sample Type

Both a tumor sample (formalin-fixed paraffin-embedded) and peripheral blood (EDTA) are necessary for MSI testing.

Limitations

It is important that the tumor specimen submitted for testing is not contaminated with large numbers of normal cells. Large percentages of contaminating normal cells in a tumor sample can result in a false-negative result.

- ***BRAF* V600E mutation testing**
  - Mutations in the *BRAF* gene are detected in 10% of colon cancers.
  - The *BRAF* gene produces a serine-threonine kinase that is constitutively activated when mutated.

- The overwhelming majority (90%) of *BRAF* gene mutations involve a change of valine to glutamic acid at codon 600 (V600E).
- The *BRAF* V600E mutation is clinically relevant because it is generally only observed in sporadic tumors. Therefore, the presence of a *BRAF* V600E mutation indicates that a diagnosis of HNPCC is unlikely. In addition, it has been reported that the presence of a *BRAF* V600E mutation indicates that the patient is not likely to respond favorably to epidermal growth factor receptor (EGFR) inhibitor therapy. BRAF kinase inhibitors are currently the subject of several clinical trials.
- Molecular testing for the *BRAF* V600E mutation is typically performed by PCR followed by pyrosequencing. This method allows for the detection of the mutation in a lower number of cells than many other methods.
- *BRAF* V600E mutation testing is typically ordered on patients with a normal IHC result but abnormal MSI PCR results or in patients with abnormal MLH1 IHC results (discussed in more detail below).

Sample Type

Formalin-fixed paraffin-embedded tumor tissue.

Limitations

Large amounts of contaminating normal tissue can result in a false-negative result. In addition, a negative result does not exclude the present of a disease-causing *BRAF* gene mutation not detected by the assay.

- ***MLH1* promoter methylation testing**
  - Methylation is a chemical modification of DNA that occurs on CG dinucleotide sequences known as CpG islands. Hypermethylation of CpG islands in the promoter of a gene turns off gene transcription effectively eliminating the production of any gene product.

- The *MLH1* gene is a MMR gene commonly mutated in patients with HNPCC. However, biallelic hypermethylation of the *MLH1* promoter is observed in a large proportion of sporadic colon cancer.
- *MLH1* promoter hypermethylation results in MSI and must be ruled out in order to determine if the MSI is sporadic or related to HNPCC.
- *MLH1* promoter hypermethylation in sporadic tumors is strongly associated with the *BRAF* V600E mutation. Therefore, individuals with an abnormal MLH1 IHC result should be tested for the *BRAF* V600E mutation prior to methylation testing to determine if the loss of MLH1 is likely to be sporadic in nature.
- Molecular testing for *MLH1* promoter methylation is typically performed by some variation of methylation specific PCR. Methylation specific PCR is a method that takes advantage of changes that occur in chemically (sodium bisulfite) treated DNA in the absence of methylation. Specifically:
  - When unmethylated DNA is treated by sodium bisulfite, all cytosines are converted to uracil. This conversion does not occur when DNA is methylated.
  - PCR primers are then designed that recognize the wild-type (methylated) sequence as well as that expected if the DNA is unmethylated and has undergone sodium bisulfite conversion. Amplification of the methylation specific primers only occurs when methylation is present in the area of interest.
  - Several different variations of methylation specific PCR exist and are commonly used in the clinical molecular diagnostics laboratory.
  - Methylation-specific PCR is highly sensitive and can be quantitative.
- Patients that are positive for *MLH1* promoter methylation but negative for the *BRAF* V600E mutation should be tested for a germline *MHL1* gene mutation.

Sample Type

Formalin-fixed paraffin-embedded tumor tissue.

Limitations

Large amounts of contaminating normal tissue can result in a false-negative result. In addition, methylation in areas other than the gene promoter will not be detected.

- **MMR gene testing**
  - Four MMR genes are responsible for the development of HNPCC:
    1. *MLH1*
    2. *MSH2*
    3. *MSH6*
    4. *PMS2*
  - Approximately 90% of HNPCC cases are caused by mutations in either the *MLH1* or *MSH2* genes.
  - Mutations in the *MSH6* gene account approximately 10% of HNPCC cases.
  - Mutations in the *PMS2* gene are rare.
  - Molecular testing for mutations in these genes is performed by Sanger sequencing of the entire coding regions and intron/exon boundaries for each gene. In addition, MLPA is performed to detect large deletions/duplications in patients that are not found to have a mutation by sequencing analysis.
  - Molecular testing for patients with HNPCC should ideally be directed by abnormal IHC results (e.g., absent PMS2 indicates that *PMS2* full gene analysis should be performed). However, it should be noted that the loss of PMS2 on IHC may be associated with an MLH1 mutation in some (rare) cases.
  - In some cases, IHC testing is normal and the family history and MSI testing dictates that a diagnosis of HNPCC is likely. In

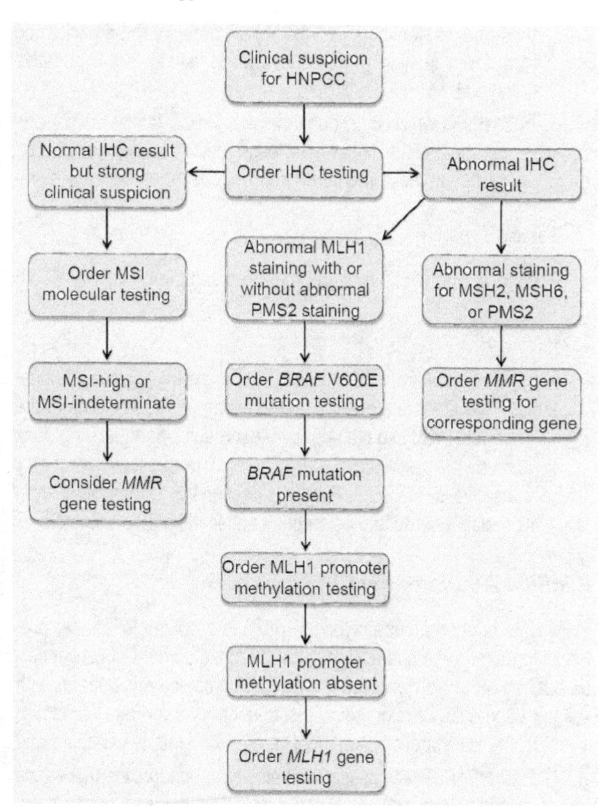

Figure 5. Molecular testing algorithm for a suspected diagnosis of HNPCC. HNPCC, hereditary nonpolyposis colorectal cancer; IHC, immunohistochemistry; MSI, microsatellite instability. *Adapted from Samowitz WS (2007–current). Lynch syndrome – hereditary nonpolyposis colorectal cancer (HNPCC). In: Jackson BR, ed. ARUP Consult: The physician's guide to laboratory test selection and interpretation. http://www.arupconsult.com (Accessed on May 23, 2012).*

those cases, testing for all four MMR genes can be performed in a tiered fashion (e.g., *MLH1*/*MSH2* testing and if negative followed by *MSH6* testing).

- In cases in which individuals have a strong family history and no personal history of cancer testing for all four genes can be performed in the same tiered manner described above.

Sample Type

Peripheral blood (EDTA) is the preferred sample type.

Limitations

A negative result from the analysis of a single MMR gene does not exclude the presence of a disease-causing mutation in one of the other genes. In addition, a negative test result by sequencing does not exclude the presence of a large deletion or duplication that may be causative of HNPCC. Finally, mutations in genes other than MMR causing hereditary colorectal cancer will not be detected.

Familial Adenomatous Polyposis

Familial adenomatous polyposis (FAP) is a colorectal cancer predisposition syndrome characterized by the development of precancerous adenomatous polyps in the colon. Without colectomy, patients with FAP will eventually develop colorectal cancer (average age of diagnosis is 39). Clinically, patients with FAP are divided into two categories:

1. Classical FAP: These patients develop from hundreds to thousands of adenomatous polyps. Without colectomy, these patients will eventually develop cancer (average age of diagnosis: 39 y).
2. Attenuated FAP: These patients develop fewer than 100 polyps and typically have later onset of colorectal cancer than those with classical FAP.

Tests

FAP (and attenuated FAP) have both traditionally been associated with autosomal dominant mutations in the *APC* gene. However, autosomal recessive mutations in the *MUTYH* (formerly *MYH*) gene have been recently implicated in these disorders (albeit more commonly in attenuated FAP). In this section molecular testing for both *APC* and *MUTYH* gene mutations are discussed in detail.

A testing algorithm for a suspected diagnosis of FAP (or attenuated FAP) in an individual of European Caucasian descent is shown in Figure 6.

- ***APC* full gene testing**
  - The *APC* gene produces a tumor suppressor protein involved in maintaining proper apoptosis and cellular growth control. Mutations in this gene typically result in the truncation of the protein product and subsequent loss of function.
  - 95% of individuals with a clinical diagnosis of classical FAP and 30% of those with attenuated FAP have a detectable mutation in the *APC* gene.
  - Mutations in the *APC* gene are located throughout the gene. Therefore, molecular testing for *APC* gene mutations involves Sanger sequencing of the entire coding region along with intron/exon boundaries.
  - Individuals that are negative for a mutation by Sanger sequencing are typically reflexed to testing for large deletions/duplications by MLPA analysis. It is estimated that 8%–12% of disease-causing *APC* gene mutations are detected by this form of testing.
  - Protein truncation testing, which was originally used to detect mutations in the *APC* gene, is no longer commonplace in clinical molecular laboratories.

Sample Type

Peripheral blood (EDTA) is the preferred sample type.

Limitations

This test will not detect causative mutations in genes other than *APC* that cause FAP and attenuated FAP. In addition, other genetic causes of hereditary colorectal cancer will not be detected by *APC* gene testing.

- ***MUTYH* targeted mutation testing**
  - The *MUTYH* gene produces a protein involved in DNA damage repair.
  - Mutations in *MUTYH* gene are reportedly identified in approximately 30% of (*APC* mutation negative) patients with FAP and 40%–55% of (*APC* mutation negative) patients with attenuated FAP.
  - Two mutations in the *MUTYH* gene, c.495A>G (p.Tyr165Cys) and c.1145G>A (p.Gly382Asp), comprise 75%–85% of disease-causing mutations in individuals of European Caucasian descent.
  - Targeted testing for these two common *MUTYH* gene mutations is routinely offered in clinical molecular laboratories as a quick and inexpensive alternative to *MUTYH* full gene sequencing.
  - Molecular testing for the two common mutations in the *MUTYH* gene typically involves allele specific PCR followed by capillary electrophoresis (or any number of detection methodologies).
  - Homozygosity for either of these mutations (or compound heterozygosity for both) confirms a diagnosis of *MUTYH*-associated FAP.
  - The detection of a single mutation indicates that *MUTYH* full gene testing is necessary in order to confirm a diagnosis (or rule it out).
  - This test is not indicated in individuals not of European Caucasian descent.

Sample Type

Peripheral blood (EDTA) is the preferred sample type.

Limitations

A negative result from targeted *MUTYH* testing does not exclude the presence of another *MUTYH* gene mutation not detected by the targeted testing. In addition, this testing does not rule out the presence of a disease-causing mutation in the *APC* gene.

- ***MUTYH* full gene sequencing**
  - Clinically indicated individuals (with no detectable *APC* gene mutation) not of European Caucasian descent and those found to have a single *MUTYH* gene mutation by targeted should undergo *MUTYH* full gene sequencing.
  - *MUTYH* full gene sequencing involves the Sanger sequencing of coding regions and intron/exon boundaries of the *MUTYH* gene.
  - The detection of two *MUTYH* gene mutations confirms a diagnosis of *MUTYH*-associated FAP/attenuated FAP.

Sample Type

Peripheral blood (EDTA) is the preferred sample type.

Limitations

The detection of two gene mutations does not confirm that these mutations are on opposite chromosomes and additional (familial) testing may be necessary to confirm their chromosomal origin. A negative result from *MUTYH* testing does not exclude the presence of an *APC* gene mutation or a mutation in other genes related to hereditary colorectal cancer.

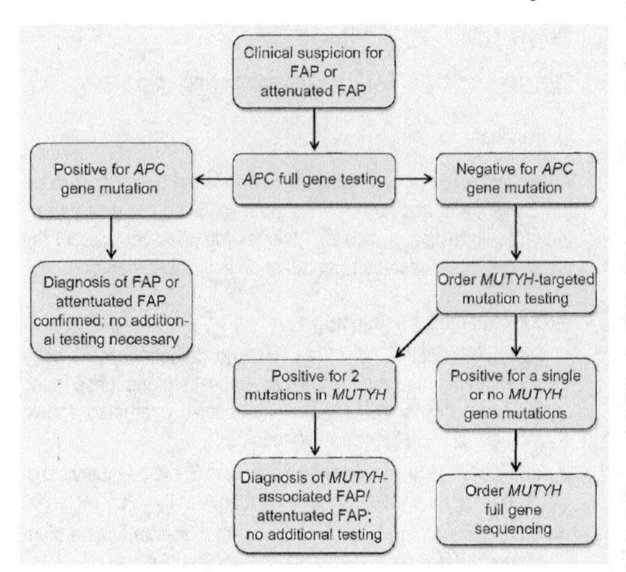

Figure 6. Molecular testing algorithm for a suspected diagnosis of FAP (or attenuated FAP) in an individual of European Caucasian descent. FAP, familial adenomatous polyposis.

Juvenile Polyposis Syndrome

Juvenile polyposis syndrome (JPS) is a disorder of the GI tract characterized by the development of "juvenile" polyps in the stomach, small intestine, colon, and rectum. While the majority of these polyps are benign, some may become malignant. Therefore, these patients are at an increased risk for colorectal cancer and other cancers of the GI tract.

Characteristics

- Hamartomatous polyps in GI tract (number of polyps varies)
- Anemia
- Abdominal pain
- Rectal bleeding

Tests

JPS is inherited in autosomal dominant manner and is caused by mutations in the *BMPR1* and *SMAD4* genes. Molecular testing for JPS is therefore focused on confirming a diagnosis through the detection of pathogenic mutations in these genes. In this section, molecular testing for *BMPR1* and *SMAD4* gene mutations is discussed in detail.

- ***BMPR1A* full gene analysis**
  - The *BMPR1A* gene produces a protein that acts as a type I receptor in the bone morphogenetic protein (BMP) pathway.
  - Mutations in the *BMPR1A* gene are identified in approximately 20% of patients with JPS.
  - Mutations in the *BMPR1A* gene are located throughout the gene. Therefore, molecular testing for mutations in the *BMPR1A* gene involves Sanger sequencing of all coding regions and intron/exon boundaries. Individuals negative for *BMPR1A* pathogenic mutations detected by Sanger sequencing are routinely tested for large deletions in this gene by multiplex ligation-dependent probe amplification. An estimated 6% of patients with JPS harbor large deletions in the *BMPR1A* gene.
  - The detection of single pathogenic mutation in the *BMPR1A* gene confirms a diagnosis of JPS. Individuals with clinical symptoms of JPS with no identifiable *BMPR1A* mutation should be tested for mutations in the *SMAD4* gene.

Sample Type

Peripheral blood (EDTA) is the preferred sample type.

Limitations

The lack of detection of pathogenic *BMPR1A* mutation does not rule out the possibility of a JPS disease-causing mutation in the *SMAD4* gene (or in other yet to be identified causative genes).

- ***SMAD4* full gene analysis**
  - The *SMAD4* gene produces a protein that acts as an intracellular mediator in the tumor growth factor (TGF)-β signaling pathway.
  - Mutations in the *SMAD4* gene are identified in approximately 20% of patients with JPS.
  - Interestingly, mutations in this gene also cause hereditary hemorrhagic telangiectasia (HHT), an autosomal dominant vascular malformation syndrome, in 15%–22% of cases. In these individuals a combined JPS/HHT syndrome is observed. As such, individuals found to have a pathogenic mutation in the *SMAD4* gene must also be monitored closely for symptoms of HHT.
  - Mutations in the *SMAD4* gene are located throughout the gene. Therefore, molecular testing for mutations in *SMAD4* gene involves Sanger sequencing of all coding regions and intron/exon boundaries. Individuals negative for *SMAD4* pathogenic mutations detected by Sanger sequencing are routinely tested for large deletions in this gene by multiplex ligation-dependent probe amplification. An estimated 4% of patients with JPS harbor large deletions in the *SMAD4* gene.
  - The detection of single pathogenic mutation in the *SMAD4* gene confirms a diagnosis of JPS. Individuals with clinical symptoms of JPS with no identifiable *SMAD4* mutation should be tested for mutations in the *BMPR1A* gene.

Sample Type

Peripheral blood (EDTA) is the preferred sample type.

Limitations

The lack of detection of a pathogenic *SMAD4* mutation does not rule out the possibility of a JPS disease-causing mutation in the *BMPR1A* gene (or in other yet to be identified causative genes).

Colorectal Cancer Drug Response

Variability in the response to chemotherapeutic drugs is a major hurdle for determining treatment modalities in many forms of cancer. Over the past several years, several molecular alterations have been identified that play a key role in determining how well a patient is likely to respond to specific types of chemotherapy.

Tests

Several relevant molecular markers are routinely used by clinicians to aid in determining drug response in colorectal cancer patients. In this section, a few of the most commonly ordered molecular tests are discussed in detail.

- ***KRAS* targeted mutation analysis**
  - The *KRAS* gene produces a protein that is important in many molecular signaling pathways.
  - Mutation of the *KRAS* gene occurs in 30%–50% of colorectal cancers.
  - Clinically, the presence of certain *KRAS* gene mutations results in an inhibited response to anti-EGFR therapy (i.e., cetuximab).
  - Mutations at *KRAS* codons 12, 13, and 61 represent the majority of those that result in decreased response to anti-EGFR therapy but other (less common) mutations have been described.

- Molecular testing for *KRAS* mutations is typically targeted to the three most commonly mutated codons.
- Several molecular methodologies exist for the detection of *KRAS* mutations but the most commonly used is pyrosequencing. Pyrosequencing, a sequencing by synthesis method, allows for the detection of a mutation in as few as 5% of cells making it ideal for mutation detection in tumor samples that commonly have large amounts of contaminating normal tissue. In addition, pyrosequencing allows for the detection of any number of possible nucleotide substitutions at the codons of interest (a distinct advantage over other possible methodologies such as AS PCR).
- The detection of a mutation in the *KRAS* gene indicates that the patient is less likely to respond to anti-EGFR therapy and other treatment modalities should be pursued.
- Individuals that are negative for *KRAS* gene mutations should be considered for *BRAF* V600E mutation testing (see below).

Sample Type

Formalin-fixed paraffin-embedded tumor tissue.

Limitations

A negative result does not rule out the presence of a *KRAS* mutation not detected by targeted testing. In addition, a false-negative result can be obtained if fewer than 5% of cells in a tested sample are from tumor.

- ***BRAF* V600E mutation testing**
  - Mutations in the *BRAF* gene are detected in 10% of colon cancers.
  - The *BRAF* gene produces a serine-threonine kinase constitutively activated when mutated.
  - The overwhelming majority (90%) of *BRAF* gene mutations involves a change of valine to glutamic acid at codon 600 (V600E).

- Similar to mutations in the *KRAS* gene (described above), the presence of the *BRAF* V600E mutation has been associated with inhibited response to anti-EGFR therapy.
- Molecular testing for the *BRAF* V600E mutation in colorectal cancer cases for determination of therapeutic response is typically performed by pyrosequencing. This allows for the detection of the *BRAF* V600E in as few as 5% of the cells in a sample.
- The detection of the *BRAF* V600E mutation indicates that the patient is unlikely to respond to anti-EGFR therapy and other treatment modalities should be pursued.

Sample Type

Formalin-fixed paraffin-embedded tumor tissue.

Limitations

A negative result does not rule out the presence of a *BRAF* mutation not detected by targeted testing. In addition, a false-negative result can be obtained if fewer than 5% of cells in a tested sample are from tumor.

- ***UGT1A1* genotyping**
  - Irinotecan is a drug used for the treatment of advanced stage colorectal cancer.
  - 20%–35% of individuals treated with irinotecan experience severe side effects.
  - The *UGT1A1* gene produces an enzyme responsible for the clearance of certain drugs (including irinotecan) and other metabolic compounds.
  - Variation in the length of a TA dinucleotide repeat in the promoter of the *UGT1A1* gene have been associated with decreased gene expression and increased risk for irinotecan-related drug toxicity. Six TA repeats in the *UGT1A1* promoter is considered to be the wild-type allele (known as *1) while a

repeat number of 7 (known as the *28 allele) is the most common cause of irinotecan-related drug toxicity.

- Molecular genotyping of the *UGT1A1* promoter is performed by PCR targeted for this region followed by capillary electrophoresis.
- The detection of a single copy of the *28 allele may indicate that an irinotecan dosage reduction may be recommended but should be based on clinical findings. However, the detection of two copies of the *28 allele indicates that a reduction in irinotecan dosage is recommended.

Sample Type

Peripheral blood (EDTA) is the preferred sample type.

Limitations

A negative result does not rule out other factors that may influence irinotecan toxicity.

Gastrointestinal Stromal Tumor

GI stromal tumors (GISTs) are the most commonly observed mesenchymal tumors of the GI tract. GISTs are usually observed in the stomach but can occur in anywhere in the GI tract. As many as 30% of these tumors cause no symptoms and go undiagnosed. However, these tumors can be malignant and aggressive requiring medical intervention.

Characteristics

- GI bleeding
- Abdominal pain
- Abdominal swelling
- Loss of appetite
- Unexplained weight loss

Tests

Somatic (and inherited) gene mutations occur in the vast majority of adult GIST cases. These mutations occur in the *KIT* and *PDGFRA* genes and can affect treatment modalities and the aggressiveness of the neoplasm. These tests are not performed to confirm a diagnosis of GIST but are used to aid clinicians in directing therapy. Molecular testing for *KIT* and *PDGFRA* mutations as they relate to GISTs are discussed in detail below.

- ***KIT* mutation testing**
  - The *KIT* gene produces a receptor tyrosine kinase that is constitutively activated in GISTs.
  - Primary activating mutations (those detected prior to treatment) in the *KIT* gene occurs in 60%–80% of GISTs.
  - 70%–75% of GIST mutations in *KIT* occur in exon 11 of the gene. The vast majority of these mutations are in-frame deletions but missense mutations have also been described.
  - GIST *KIT* mutations are also relatively common in exon 9 (10%–20% of *KIT* mutations). They have also been reported to occur (rarely) in exons 13 and 17.
  - Molecular testing for *KIT* mutations typically involves Sanger sequencing for exons 9 and 11 of the gene. This methodology will detect both deletions and missense mutations. If patients are negative mutations in either exon 9 or 11 then testing for rare mutations (those in exons 13 and 17) can be pursued.
  - The type of *KIT* mutation detected provides information on both the behavior of the tumor as well as treatment.
    - *KIT* exon 9 mutations are associated with very aggressive tumors and may require the use of higher doses of imatinib or alternative TKI therapy (e.g., sunitinib).
    - *KIT* exon 11 mutations are associated with an overall favorable response to TKI (imatinib) therapy and better overall survival.

    - *KIT* exon 13 and 17 mutations typically do not respond well to TKI therapy.
  - Secondary *KIT* mutations (those occurring after initiation of treatment) are often observed in patients undergoing TKI therapy. These mutations often result in resistance to therapy and may indicate that alternative treatment modalities should be pursued.
  - Patients with GIST who are negative for mutations in the *KIT* gene should undergo testing for *PDGFRA* mutation testing.

Sample Type

Formalin-fixed paraffin-embedded tumor tissue.

Limitations

A positive result for a *KIT* mutation does not confirm a diagnosis of GIST as these mutations are observed in several other tumor types. Large amounts of contaminating normal tissue in the tested sample can result in a false-negative result. Other factors can contribute to TKI response/resistance that will not be detected by this assay.

- ***PDGFRA* mutation testing**
  - The *PDGFRA* gene produces a receptor tyrosine kinase that is constitutively activated in GISTs.
  - Activating mutations in the *PDGFRA* gene occur in 5%–10% of patients with GIST. These mutations are not observed in the presence of *KIT* gene mutations.
  - The most common *PDGRFA* gene mutations are missense mutations that occur in exon 18 of the gene. *PDGFRA* mutations are observed in exon 12 and rarely observed in exon 14.
  - Molecular testing for *PDGFRA* gene mutations in GIST patients typically involves Sanger sequencing of exons 12 and 18. This methodology will detect disease-causing mutations located

anywhere in these exons. Molecular testing for rare mutations in *PDGFRA* exon 14 is not commonly available.

- *PDGFRA* mutation positive GIST tumors are generally less aggressive than those with *KIT* mutations and respond well to TKI therapy (e.g., imatinib).
- Certain *PDGFRA* mutations (e.g., p.Asp842Val) have demonstrated resistance to TKI therapy indicating that other therapeutic modalities should be pursued in the presence of these mutations.

Sample Type

Formalin-fixed paraffin-embedded tumor tissue.

Limitations

Large amounts of contaminating normal tissue in the tested sample can result in a false-negative result. Other factors can contribute to TKI response/resistance that will not be detected by this assay.

Oligodendroglioma

Oligodendroglioma is one of the more common neoplasms observed in the central nervous system (CNS). However, unlike many CNS tumors, oligodendrogliomas are noteworthy as these tumors show a favorable response to chemotherapy and better overall clinical outcome. Histologically, oligodendrogliomas are very similar to astrocytomas, a tumor with a less favorable response to therapy, making them difficult to diagnosis using traditional methodologies (such as microscopy).

Tests

Molecular testing is an important tool that can aid in differentiating between the histologically similar oligodendroglioma and astrocytoma. Not only can this testing help determine a diagnosis but by default helps determine treatment. Molecular testing on these CNS tumor samples is described in detail below.

- **1p/19q FISH analysis**
  - 70%–90% of oligodendrogliomas exhibit loss of heterozygosity (LOH) of both chromosomes 1p and 19q. Loss of these chromosomes is very rare in astrocytoma.
  - Several studies have demonstrated that oligodendrogliomas exhibiting this pattern of LOH are more likely to respond to chemotherapy. In addition, patients exhibiting LOH have a better overall survival.
  - Although many methodologies exist, the most commonly performed molecular test for LOH of 1p and 19q in oligodendrogliomas is FISH. This method uses probes specific for the chromosomal regions of interest and can quickly (and inexpensively) identify LOH in tumor samples. In addition, this methodology can also identify duplications of 19q, which are occasionally observed in astrocytomas but not oligodendrogliomas.
  - The presence of LOH on both chromosome 1p and 19q supports a diagnosis of oligodendroglioma. In addition, the loss of chromosome 1p only is also suggestive of a diagnosis of oligodendroglioma.

Sample Type

Formalin-fixed Paraffin-embedded tumor tissue.

Limitations

As not all tumors have the same pattern of LOH, a negative result does not rule out a diagnosis of oligodendroglioma.

Suggested Reading

American Cancer Society. http://www.cancer.org (Accessed May 13, 2013).

Andersson J, Bumming P, Meis-Kindblom JM, et al. Gastrointestinal stromal tumors with KIT exon 11 deletions are associated with poor prognosis. Gastroenterology 2006;130:1573–81.

Antonescu CR, Besmer P, Guo T, et al. Acquired resistance to imatinib response in gastrointestinal stromal tumor occurs through secondary gene mutation. Clin Cancer Res 2005;11:4182–90.

Aretz S, Uhlhaas S, Goergens H, et al. MUTYH-associated polyposis: 70 of 71 patients with biallelic mutations present with an attenuated or atypical phenotype. In J Cancer 2006;119:807–14.

Baccarani M, Saglio G, Goldman J, et al. Evolving concepts in the management of chronic myeloid leukemia: recommendations from an expert panel on behalf of the European LeukemiaNet. Blood 2006;108:1809–20.

Bayraktar UD, Baraktar S, Rocha-Lima S. Molecular basis and management of gastrointestinal stromal tumors. World J Gastroentrol 2010;16:2726–34.

Bedeir A, Krasinskas AM. Molecular diagnostics of colorectal cancer. Arch Pathol Lab Med 2011;135:578–87.

Cairncross JG, Ueki K, Zlatescu MC, et al. Specific genetic predictors of chemotherapeutic response and survival in patients with anaplastic oligodendrogliomas. J Natl Cancer Inst 1998;90:1473–9.

Carlson RW, Allred DC, Anderson BO, et al. Breast cancer: clinical practice guidelines in oncology. J Natl Canc Netw 2009;7:122–92.

Calva-Cerqueira D, Chinnathambi S, Pechman B, et al. The rate of germline mutations and large deletions of SMAD4 and BMPR1A in juvenile polyposis. Clin Genet 2009;75:79–85.

Corless CL, Schroeer A, Griffith D, et al. PDGFRA mutations in gastrointestinal stromal tumors: frequency, specrtrum and in vitro sensitivity to imatinib. J Clin Oncol 2005;23:5357–64.

Croitoru ME, Cleary SP, Di Nicola N, et al. Association between biallelic and monoallelic germline MYH gene mutations and colorectal cancer risk. J Natl Cancer Inst 2004;96:1631–4.

Cushman-Vokoun AM. Somatic alterations and targeted therapy. In: Best DH, Swensen JJ, eds. Molecular Genetics and Personalized Medicine. New York: Humana, 2012:51–101.

Davies H, Bignell GR, Cox C, et al. Mutations of the BRAF gene in human cancer. Nature 2002;417:949–54.

de The H, Chen Z. Acute promyelocytic leukaemia: novel insights into the mechanisms of cure. Nat Rev Cancer 2010;10:775–83.

Debiec-Rychter M, Sciot R, Le Cesne A, et al. KIT mutations and dose selection for imatinib in patients with advanced gastrointestinal stromal tumours. Eur J Cancer 2006;42:1093–103.

Di Fiore F, Blanchard F, Charbonnier F, et al. Clinical relevance of KRAS mutation detection in metastatic colorectal cancer treated by Cetuximab plus chemotherapy. Br J Cancer 2007;96:1166–9.

Di Nicolantonio F, Martini M, Molinari F, et al. Wild-type BRAF is required for response to panitumumab or cetuximab in metastatic colorectal cancer. J Clin Oncol 2008;26:5705–12.

Duffy MJ, O'Donovan N, Crown J. Use of molecular markers for predicting therapy response in cancer patients. Cancer Treat Rev 2011;37:151–9.

Filipe B, Baltazar C, Albuquerque C, et al. APC or MUTYH mutations account for the majority of clinically well-characterized families with FAP and AFAP phenotype and patients more than 30 adenomas. Clin Genet 2009;76:242–55.

Finkelstein SD, Sayegh R, Christensen S, Swalsky PA. Genotypic classification of colorectal adenocarcinoma: Biologic behavior correlates with K-ras-2 mutation type. Cancer 1993;71:3827–38.

Fleeman N, Martin Saborido C, Payne K, et al. The clinical effectiveness and cost-effectiveness of genotyping for CYP2D6 for the management of women with breast cancer treated with tamoxifen: a systematic review. Health Technol Assess 2011;15:1–102.

Foran JM. New prognostic markers in acute myeloid leukemia: perspective from the clinic. Hematology 2010;2010:47–55.

Gallione CJ, Repetto GM, Legius E, et al. A combined syndrome of juvenile polyposis and hereditary haemorrhagic telangiectasia associated with mutations in the MADH4 (SMAD4). Lancet 2004;363:852–9.

Geirsbach KB, Samowitz WS. Microsatellite instability and colorectal cancer. Arch Pathol Lab Med 2011;135:1269–77.

Gismondi V, Meta M, Bonelli L, et al. Prevalence of the Y165C, G382D and 1395delGGA germline mutations of the MYH gene in Italian patients with adenomatous polyposis coli and colorectal adenomas. Int J Cancer 2004;109:680–4.

Hahn KK, Wolff JJ, Kolesar JM. Pharmacogenetics and irinotecan therapy. Am J Health Syst Pharm 2006;63:2211–7.

Hampel H. Point: justification for Lynch Syndrome screening among all patients with newly diagnosed colorectal cancer. J Natl Compr Canc Netw 2010;8:597–601.

Hartmann C, Mueller W, von Deimling A. Pathology and molecular genetics of oligodendroglial tumors. J Mol Med 2004;82:638–55.

Harrison CN, Bareford, D, Butt N, et al. Guideline for investigation and management of adults and children presenting with a thrombocytosis. Br J Haematol 2010;149:352–75.

Heinrich MC, Corless CL, Duensing A, et al. PDGFRA activating mutations in gastrointestinal stromal tumors. Science 2003;299:708–10.

Hoskins JM, Carey LA, McLeod HL. CYP2D6 and tamoxifen: DNA matters in breast cancer. Nat Rev Cancer 2009;9:576–86.

Hughes T, Deininger M, Hochhaus A., et al. Monitoring CML patients responding to treatment with tyrosine kinase inhibitors: review and recommendations for harmonizing current methodoly for detecting BCR-ABL transcripts and kinase domain mutations and for expressing results. Blood 2006;108:28–37.

Ino Y, Zlatescu MC, Sasaki H, et al. Long survival and therapeutic responses in patients with histologically disparate high-grade gliomas demonstrating 1p loss. J Neurosurg 2000;92:983–90.

Innocenti F, Undevia SD, Iyer L, et al. Genetic variants in the UDP-glucuronosyltransferase 1A1 gene predict the risk of severe neutropenia of irinotecan. J Clin Oncol 2004;22:1382–8.

Isidro G, Laranjeira F, Pires A, et al. Germline MUTYH (MYH) mutations in Portuguese individuals with multiple colorectal adenomas. Hum Mutat 2004;24:353–4.

Jabbour EJ, Estey E, Kantarjian HM. Adult acute myeloid leukemia. Mayo Clin Proc 2006;81:247–60.

Jasperson KW, Burt RW. APC-associated polyposis conditions. 1998 Dec 18 [Updated 2011 Oct 27]. In: Pagon RA, Bird TD, Dolan CR, et al., editors. GeneReviews™ [Internet]. Seattle (WA): University of Washington, Seattle; 1993-. Available from: http://www.ncbi.nlm.nih.gov/books/NBK1345/ (Accessed May 13, 2013).

Jones D, Kamel-Reid S, Bahler D, et al. Laboratory practice guidelines for detecting and reporting BCR-ABL drug resistance mutations in chronic myelogenous leukemia and acute lymphoblastic leukemia: a report of the Association for Molecular Pathology. J Mol Diag 2009;11:4–11.

Klco JM, Vij R, Kreisel FH, et al. Molecular pathology of myeloproliferative neoplasms. Am J Clin Pathol 2010;133:602–15.

Kohlmann W, Gruber SB. Lynch syndrome. 2004 Feb 5 [Updated 2012 Sep 20]. In: Pagon RA, Bird TD, Dolan CR, et al., editors. GeneReviews™ [Internet]. Seattle (WA): University of Washington, Seattle; 1993-. Available from: http://www.ncbi.nlm.nih.gov/books/NBK1211/ (Accessed May 13, 2013).

Kralovics R, Passamonti F, Buser AS, et al. A gain of function mutation of JAK2 in myeloproliferative disorders. N Engl J Med 2005; 352:1779–90.

Larsen Haidle J, Howe JR. Juvenile polyposis syndrome. 2003 May 13 [Updated 2011 Sep 29]. In: Pagon RA, Bird TD, Dolan CR, et al., editors. GeneReviews™ [Internet]. Seattle (WA): University of Washington, Seattle; 1993-. Available from: http://www.ncbi.nlm.nih.gov/books/NBK1469/ (Accessed May 13, 2013).

Lasota J, Miettinen M. Clinical significance of oncogenic KIT and PDGFR mutations in gastrointestinal stromal tumours. Histopathology 2008;53:245–66.

Laurini JA, Carter JE. Gastrointestinal stromal tumors: a review of the literature. Arch Pathol Lab Med 2010;134:134–41.

Loupakis F, Ruzzo A, Cremolini C, et al. KRAS codon 61, 146 and BRAF mutations predict resistance to cetuximab plus irinotecan in KRAS codon 12 and 13 wild-type metastatic colorectal cancer. Br J Cancer 2009;101:715–21.

Marcuello E, Altes A, Menovo A, et al. UGT1A1 gene variations and irinotecan treatment in patients with metastatic colorectal cancer. Br J Cancer 2004;91:678–82.

Marcucci G, Haferlach T, Dohner H. Molecular genetics of adult acute myeloid leukemia: Prognostic and therapeutic implications. J Clin Oncol 2011;29:475–86.

Martin-Broto J, Rubio L, Alemany R, Lopez-Guerrero JA. Clinical implications of KIT and PDGFRA genotyping in GIST. Clin Transl Oncol 2010;12670–6.

Oliveira C, Velho S, Moutinho C, et al. KRAS and BRAF oncogenic mutations in MSS colorectal carcinoma progression. Oncogene 2007;26:158–63.

Paz-Priel I, Friedman AD. C/EBPα dysregulation in AML and ALL. Crit Rev Oncog 2011;16:93–102.

Petrucelli N, Daly MB, Feldman GL. BRCA1 and BRCA2 Hereditary breast and ovarian cancer. 1998 Sep 4 [Updated 2011 Jan 20]. In: Pagon RA, Bird TD, Dolan CR, et al., editors. GeneReviews™ [Internet]. Seattle (WA): University of Washington, Seattle; 1993-. Available from: http://www.ncbi.nlm.nih.gov/books/NBK1247/ (Accessed May 13, 2013).

Pietra D, Li S, Brisci A, et al. Somatic mutations of JAK2 exon 12 in patients with JAK2 (V617F)-negative myeloproliferative disorders. Blood 2008;111:1686–9.

Quintas-Cardama A, Cortes J. Molecular biology of bcr-abl1-positive chronic myeloid leukemia. Blood 2009;113:1619–30.

Rouits E, Boisdron-Celle M, Dumont A, et al. Relevance of different UGT1A1 polymorphisms in irinotecan-induced toxicity: a molecular and clinical study of 75 patients. Clin Cancer Res 2004;10:515–9.

Samowitz WS. Lynch syndrome – hereditary nonpolyposis colorectal cancer (HNPCC). In: Jackson BR, ed. ARUP Consult. The Physician's Guide to Laboratory Test Selection and Interpretation. http://arupconsult.com/Topics/LynchSyndrome.html (Accessed May 13, 2013).

Scott LM, Tong W, Levine RL, et al. JAK2 exon 12 mutations in polycythemia vera and idiopathic erythrocytosis. N Engl J Med 2007;356:459–68.

Sharma SG, Gulley ML. BRAF mutation testing in colorectal cancer. Arch Pathol Lab Med 2010;134:1225–8.

Singh MS, Francis PA, Michael M. Tamoxifen, cytochrome P450 genes and breast cancer clinical outcomes. Breast 2011;20:111–8.

Smith JS, Alderete B, Minn Y, et al. Localization of common deletion regions on 1p and 19q in human gliomas and their association with histological subtype. Oncogene 1999;18:4144–52.

Smith JS, Perry A, Borrell TJ, et al. Alterations of chromosome arms 1p and 19q as predictors of survival in oligodendrogliomas, astrocytomas, and mixed oligoastrocytomas. J Clin Oncol 2000;18:636–45.

Stirewalt DL, Radich JP. The role of FLT3 in haematopoietic malignancies. Nat Rev Cancer 2003;3:650–65.

Tai W, Mahato R, Cheng K. The role of HER2 in cancer therapy and targeted drug delivery. J Control Release 2010;146:264–75.

Tan DSP, Marchio C, Reis-Filho JS. Hereditary breast cancer: From molecular pathology to tailored therapies. J Clin Pathol 2008;611073–82.

Tefferi A. Polycythemia and essential thrombocythemia: 2012 update on diagnosis, risk stratification, and management. Am J Hematol 2012;87:285–93.

Tefferi A. Primary myelofibrosis: 2012 update on diagnosis, risk-stratification, and management. Am J Hematol 2011;86:1018–26.

Tefferi A, Thiele J, Orazi A. Proposals and rationale for revision of the World Health Organization diagnostic criteria for polycythemia vera, essential thrombocythemia, and primary myelofibrosis: recommendations from an ad hoc international expert panel. Blood 2007;110:1092–7.

Tsiatis AC, Norris-Kirby A, Rich RG, et al. Comparison of Sanger sequencing, pyrosequencing, and melting curve analysis for the detection of KRAS mutations: diagnostic and clinical implications. J Mol Diagn 2010;12:425–32.

Umar A, Boland CR, Terdiman JP, et al. Revised Bethesda Guidelines for hereditary nonpolyposis colorectal cancer (Lynch Syndrome) and microsatellite instability. J Natl Cancer Inst 2004;96:261–8.

van der Groep P, van der Wall E, van Diest PJ. Pathology of hereditary breast cancer. Cell Oncol 2011;34:71–88.

Vannucchi AM. Management of myelofibrosis. Hematology Am Soc Hematol Educ Prog 2011;2011:222–30.

Vardiman JW, Harris NL, Brunning RD. The World Health Organization (WHO) classification of the myeloid neoplasms. Blood 2002;100:2292–302.

Vasen HFA, Watson P, Mecklin JP, Lynch HT. New clinical criteria for hereditary nonpolyposis colorectal cancer (HNPCC, Lynch

Syndrome) proposed by the International Collaborative Group on HNPCC. Gastroenterology 1999;116:1453–6.

Volpe G, Panuzzo C, Ulisciani S, Cilloni D. Imatinib resistance in CML. Cancer Lett 2009;274:1–9.

Walker A, Marcucci G. Impact of molecular prognostic factors in cytogenetically normal acute myeloid leukemia at diagnosis and relapse. Haematologica 2011;96:640–3.

Warner E. Breast-cancer screening. N Eng J Med 2011;365:1025–32.

Wouters BJ, Lowenberg B, Erpelinck-Verschueren CAJ, et al. Double CEBPA mutations but not single CEBPA mutations, define a subgroup of acute myeloid leukemia with distinctive gene expression profile that is uniquely associated with a favorable outcome. Blood 2009;113:3088–91.

Infectious Diseases

Molecular infectious disease testing is perhaps the oldest subdiscipline within molecular diagnostics. The powerful biochemistry of polymerase chain reaction (PCR) has allowed for the ready design and implementation of fluid-based tests that detect fastidious organisms and that can replace invasive, low-yield diagnostic procedures such as biopsy. Many of these assays have become diagnostic standards despite the lack of cleared or approved versions by the US Food and Drug Administration (FDA). In addition, the demonstration of almost-instant clinical utility and rapid adoption of PCR-based tests has spurred the development of other nucleic acid amplification chemistries (e.g., helicase-dependent amplification, loop-mediated isothermal amplification, strand displacement amplification, transcription mediated amplification) that are now used in FDA-cleared/approved pathogen detection tests.

The formats of molecular infectious diseases tests are:

- **Qualitative detection:**
  - The simplest of the three formats, these tests yield one of two possible results:
    - Pathogen nucleic acid detected or
    - Pathogen nucleic acid not detected.
  - Designed to be highly sensitive; usually most easily interpretable when the reference value is "not detected," for example in sample types that should be sterile such as cerebrospinal fluid (CSF).
  - Results should always be interpreted in conjunction with all available clinical information. Erroneous results can occur, including:
    - False positives due to contamination or residual nucleic acid in resolved infections.

        - False negatives from insufficiently sensitive tests or presence of amplification inhibitors in patient specimens.
    - Qualitative test results can be challenging to interpret in settings of:
        - Prolonged asymptomatic shedding after infection (enteroviruses, rhinoviruses in respiratory samples).
        - Asymptomatic colonization (*Clostridium difficile, Pneumocystis jiroveci*).
        - Latent infection (cytomegalovirus [CMV], Epstein Barr virus [EBV]).
- **Pathogen quantification:**
    - More complex than qualitative tests, these assays yield information regarding pathogen levels or "loads."
    - Quantitative tests should have a range of accurate measurement encompassing clinically observed pathogen concentrations.
        - Also usually designed to be as sensitive as qualitative assays, allowing qualitative (sensitive detection) and quantitative diagnostic aims to be fulfilled with a single assay.
    - Uses:
        - Prior to treatment initiation ("baseline"):
            - Indicator of disease activity (human immunodeficiency virus [HIV], hepatitis B virus [HBV]).
            - Determinant of when to start therapy (HIV, HBV).
        - During treatment (HIV, HBV, hepatitis C virus [HCV], CMV):
            - To assess therapeutic response.
            - To guide further management, as an indicator of antiviral resistance and the need to modify therapy.
- **Genotype determination:**
    - Formats:
        - Direct sequencing: Sequence "reads" of genomic fragments; analysis by comparison to known databases.

    - Single-sequence determination (Sanger, pyrosequencing); currently in diagnostic use.
    - Massively parallel sequencing (next-generation sequencing [NGS]); not yet in routine use.
  - Hybridization-based tests: Single nucleotide polymorphism (SNP) identification through fluorescence detection using probes (fluorescence resonance energy transfer [FRET], molecular beacons) or melt curve analysis.
    - Pathogen SNPs currently of limited diagnostic use.
- Uses:
  - Pathogen and host genome characterization to guide antiviral management (HCV).
  - Detection of drug-resistance mutations:
    - Primarily to antiviral drugs (HIV, CMV, HBV, influenza A virus).
    - Detection of antibacterial drug resistance is possible but is not yet in routine clinical use.
  - Pathogen identification:
    - Bacterial (especially mycobacterial) and fungal speciation; used in reference and specialty laboratories.

This section reviews the most common uses of molecular infectious disease testing. Overviews of important infections, test utility, available tests, sample types, and test limitations are provided.

Central Nervous System Infections

Molecular tests have revolutionized the diagnosis of viral central nervous system (CNS) infections. Conventional methods remain the tests of choice for bacterial and fungal CNS infections. Pertinent clinical findings in viral CNS infections are described in Table 12.

Table 12. Clinical Findings in Viral Infections of the Central Nervous System

Syndrome	*Signs/Symptoms*	*Imaging Findings*	*Cerebrospinal Fluid Findings*	*Viruses*
Encephalitis	Altered mental status Focal neurologic deficits	Hyperintensity of lesions on T2-weighted images (commonly temporal lobe in HSV-1 infection)	Glucose: normal Protein: commonly suppressed; can be minimally elevated Pleocytosis: usually moderate lymphocytosis (usually ≤ 100 cells/mm^3)	HSV-1 HSV-2 Enteroviruses VZV CMV (congenital infection, im- munocompromised hosts) HHV-6 (solid organ transplant, hematopoietic stem cell transplant)
Aseptic meningitis	Fever Headache Photo-/phono- phobia nuchal Nuchal rigidity	Usually not performed		Enteroviruses Parechoviruses HSV-2 VZV

Myelitis (anterior horn cell infection with lower motor neuron dysfunction)	Asymmetric flaccid paralysis Asymmetric areflexia	MRI: Hyperintensity of affected anterior horn on T2-weighted images	Glucose: normal Protein: commonly suppressed; can be minimally elevated Pleocytosis: usually moderate lymphocytosis (usually 100 cells/mm^3)	Enteroviruses
Myelitis (transverse myelitis presentation)	Motor, sensory, autonomic dysfunction	MRI: centrally located lesions, commonly over 3 segments; hyperintensity at periphery of lesions on T2-weighted images		CMV, VZV, HSV, EBV
Progressive multifocal leukoencephalopathy	Altered mental status Motor deficits Ataxia Visual deficits	Multifocal demyelinating lesions hyperintensity of lesions on T2-weighted images	Usually normal; can have minimally elevated protein and mild pleocytosis (< 25 lymphs/mm^3)	JCV

CMV, cytomegalovirus; EBV, Epstein Barr virus; HHV, human herpes virus; HSV, herpes simplex virus; JCV, John Cunningham virus; MRI, magnetic resonance imaging; VZV, varicella zoster virus.

Tests

Qualitative molecular tests that amplify and detect viral genome targets are the most commonly utilized detection format; a role for pathogen quantification in CSF has yet to be firmly established.

- FDA-cleared tests are available for the detection of enteroviruses in patients presenting with symptoms of aseptic meningitis.
- Currently, only laboratory-developed tests are available for the diagnosis of other CNS syndromes.

Sample Type

CSF.

Limitations (by Virus Target)

Herpes Simplex Virus (HSV)

- False-negative results can be obtained with real-time PCR tests that amplify/detect glycoprotein genes.
  - Strain-dependent sequence heterogeneity can prevent probe binding.

Enteroviruses

- Detection of different types is achieved by amplification/detection of highly conserved 5′ untranslated region.
- Assays must be carefully designed to avoid crossreactivity with rhinoviruses.

Parechoviruses

- Detection requires assays specifically designed for amplification of parechovirus genome targets.
- Not detectable by enterovirus real-time PCR assays.

Varicella Zoster Virus (VZV)

- Nucleic acid amplification tests (NAAT) are useful for CNS disease detection in primary (varicella) infection.
- Careful clinical correlation is required for CNS disease in reactivation infections (zoster) since VZV DNA is detectable in CSF in uncomplicated zoster.

Human Herpes Virus Type 6 (HHV-6)

HHV-6 has been associated with a number of diseases in solid organ transplant and hematopoietic transplant patients; strongest association is as a causative agent of encephalitis. This diagnosis should be made once other causes have been excluded. See *Infections Managed Through Nucleic Acid Testing of Blood* for additional discussion.

CSF should be tested; false-positive results can occur in blood due to viral latency in peripheral blood mononuclear cells and frequent asymptomatic reactivation in immunocompromised individuals.

False-positive results can be observed due to viral integration into the host genome, found in ~1% of population.

John Cunningham Virus (JCV)

- Biopsy remains diagnostic gold standard.
- Negative molecular test result does not rule out progressive multifocal leukoencephalopathy (PML) if diagnosis is supported by symptoms and imaging studies.
    - False-negative molecular test results can occur due to low CSF viral loads in combination with insensitive assays.

Respiratory Infections

Molecular tests have been developed for respiratory infections to speed the time to diagnosis and increase sensitivity of detection for fastidious organisms. These tests are currently used for viruses, and some fastidious bacteria. Conventional tests are still the mainstays of diagnosis for common community-acquired bacterial infections and for fungal infections in immunocompromised patients.

Viral

In immunocompetent patients, typical upper respiratory symptoms associated with these infections are fever, rhinorrhea, sore throat, cough, and myalgias. The Centers for Disease Control and Prevention defines influenza-like illness as fever ≥ 100 °F, cough, and/or sore throat. Immunocompromised patients can present with severe symptoms that progress rapidly to lower tract disease (bronchitis, bronchiolitis, and pneumonia). Infants and the elderly are predisposed to severe or progressive disease, particularly with influenza A virus and respiratory syncytial virus. The parainfluenza viruses (types 1–3) are major causes of laryngotracheobronchitis (croup). Molecular tests have increased diagnostic yield compared to conventional methods and may allow for more rapid initiation of antiviral treatment, but have not been demonstrated to alter management and outcomes such as antibiotic use and hospitalization length of stay.

Tests

Qualitative molecular tests that amplify and detect viral genome targets are the most commonly utilized detection format; a role for pathogen quantification has yet to be firmly established. A variety of single target, small multiplex, and large multiplex formats are available as FDA-cleared tests.

Molecular tests improve the time to detection and may also increase the diagnostic yield of the following viruses that are detectable by conventional methods:

- Adenoviruses
- Human metapneumovirus*
- Influenza A*

*Viruses with RNA genomes that require reverse transcription.

- Influenza B*
- Parainfluenza virus types 1–3*
- Respiratory syncytial virus*

Molecular assays are particularly useful for the detection of fastidious viruses that are either tested for only in reference laboratories (rhinoviruses*, enteroviruses*, parainfluenza virus type 4*) or are undetectable by conventional assays (coronaviruses*).

Sample Type

- Assays have been cleared for use with posterior nasopharyngeal swab submitted in viral transport medium.
- Lower tract specimens may be useful but require validation by individual laboratories:
  - Bronchial brush/wash
  - Bronchoalveolar lavage (BAL)

Limitations

- Prolonged shedding of replicating virus occurs in infants/immunocompromised patients.
- Maintenance of isolation for prevention of nosocomial transmission should be based on culture results due to prolonged nucleic acid shedding.
- Rhinovirus detection by molecular assays requires careful clinical correlation as it can be detected by amplified nucleic acid tests in asymptomatic individuals.

Bacterial

Bordetella pertussis

- Produces severe respiratory disease with significant morbidity and mortality in unvaccinated infants. Immunity after infection or

---

*Viruses with RNA genomes that require reverse transcription.

vaccination wanes after ~10 y, creating a reservoir of susceptible individuals.

- Symptoms in unvaccinated individuals:
  - Mild upper respiratory symptoms; then
  - Paroxysmal cough (apnea in neonates); then
  - Resolving cough (bouts with lower frequency and intensity).
- Symptoms in individuals with a history of prior infection or vaccination:
  - Chronic episodic cough, often severe, with post-tussive complications such as emesis, rib fracture, and rectal prolapse.

- Pertussis is a reportable disease; positive molecular test results should be reported to public health authorities.
- Infections are treated with macrolide antibiotics (protein synthesis inhibitors); resistance is rare.
  - Close contacts should receive prophylaxis.

Tests

Qualitative nucleic acid detection in multiplex (one current cleared assay) or single-analyte (most are laboratory developed) assays. Common targets:

- *IS481*: High copy number (80–100) insertion element
  - Assays are sensitive, but can cross-react with *Bordetella holmesii*, an uncommon pathogen that can cause respiratory disease and contains this insertion element.
- Pertussis toxin promoter
  - Single copy genetic target specific for *B. pertussis*.
    - Not as sensitive as multi-copy *IS481*-based assays.

Sample Type

- Posterior nasopharyngeal swab (Dacron or Rayon; alginate will inhibit Taq polymerase).

Limitations

- False-positive results can occur as in any amplified assay.
  - Testing should only be performed in symptomatic individuals; surveillance testing in asymptomatic individuals is not recommended.
- False-negative results may occur in:
  - Tests utilizing single copy targets.
  - Previously infected or vaccinated individuals who present late (many weeks) into the course of illness.
    - Serology should be performed if clinical suspicion remains high and molecular test result is negative in this setting.

Chlamydophila pneumoniae and *Mycoplasma pneumoniae*

- Cause bronchitis. Patients present with chronic cough.
- Obligate intracellular bacteria that are well suited for molecular detection.

Tests

- Qualitative nucleic acid detection in multiplex (one current cleared assay) or single-analyte (most laboratory developed real-time PCR) assays.

Sample Type (specimens that could be validated as laboratory-developed tests)

- Nasopharyngeal swab in viral transport medium (acceptable for FDA-cleared multiplex test)
- Throat swab in viral transport medium*
- BAL*

---

*Viruses with RNA genomes that require reverse transcription.

Limitations

- *C. pneumoniae* is an uncommon cause of respiratory infection; positive molecular test results require careful clinical correlation.

Mycobacterium tuberculosis

- Patients typically present with sputum-producing chronic cough unresponsive to conventional antibiotics, and fever and weight loss. Chest radiography shows areas of consolidation (typically apical if disease is due to reactivation) with or without cavitation.
- WHO has recommended real-time PCR (Xpert MTB/RIF, not available currently in the US) for use in suspected multidrug-resistant tuberculosis and patients infected with HIV.
- Tuberculosis is a reportable disease; positive molecular test results should be reported to public health authorities.

Tests

One commercial transcription mediated amplification-based qualitative assay (Amplified MTD, Hologic Gen-probe) approved for direct detection of mycobacteria within M. tuberculosis complex (MTbC; *Mycobacterium tuberculosis*, *M. bovis, M. africanum, M. canetti*) regardless of direct stain result.

Sample Type

- Sputum.
- Lower respiratory samples (tracheal aspirate, bronchial aspirate, bronchoalveolar lavage).

Limitations

- Culture is required despite initial test results.
  - Negative result does not exclude diagnosis; culture is gold standard for sensitivity.

  - For positive results, culture is required for susceptibility testing and to distinguish which *M. tuberculosis* complex (MTbC) organism is present.
- *M. tuberculosis* direct (MTD) test detects other organisms in addition to *M. tuberculosis*; false-positive results due to detection of other MTbC organisms could occur but these infections are rare.
- MTD should not be used as test of cure.

Streptococcus (Group A)

- Indication for use in pharyngitis as initial diagnostic test or to confirm negative antigen test.

Tests

One commercial hybridization assay available (GASDirect Test, Hologic Gen-probe); sensitivity/specificity compared with culture: 91.7%/99.3%.

Sample Type

Throat swab.

Limitations

- Negative result cannot exclude infection.
- Positive result cannot distinguish between colonization and active infection.

Sexually Transmitted Infections

The advantages of molecular tests for detection of sexually transmitted infections compared with conventional tests are increased sensitivity and the availability of high throughput automated systems that facilitate screening appropriate populations, particularly for *Chlamydia trachomatis, Neisseria gonorrhoeae,* and human papillomavirus (HPV) types associated with a high risk of cervical cancer ("high-risk HPV").

C. trachomatis serovars D-K and *N. gonorrhoeae*

- These pathogens are the major causes of urethritis in males and females and cervicitis in females.
- Infections can be asymptomatic in males and females.
  - In females, asymptomatic infection can cause progressive ascending infection leading to pelvic inflammatory disease and its sequelae (infertility, ectopic pregnancy).
  - Asymptomatic screening currently recommended for females only.
    - *C. trachomatis*: All females < 25 y, older, at-risk older women (new sex partner, multiple sex partners).
    - *N. gonorrhoeae*: Females at high risk (e.g., commercial sex worker).
- Rectal and oropharyngeal infections can occur, particularly in men who have sex with men.
- Treatment for both infections with ceftriaxone (for *N. gonorrhoeae*) and azithromycin or doxycycline (for *C. trachomatis*) is recommended since coinfections are common.
- Partners (within 60 days of symptom onset or diagnosis) should be referred for evaluation and treatment to prevent re-infection and spread to others.
- Both infections are reportable diseases; positive molecular test results should be reported to public health authorities.

L1, L2, L3 (Lymphogranuloma venereum [LGV])

- These pathogens cause genital/anal ulcers, proctocolitis.
  - Diagnosis can be delayed due to symptom overlap with other entities.
  - Transmitted during oral and receptive anal intercourse.

- LGV is usually a clinical diagnosis in the setting of an ulcer disease with positive *C. trachomatis* NAAT.
    - Confirmation of L1–L3 serovars requires a separate test.
- Requires longer treatment than non-LGV disease.
    - Definitive diagnosis of LGV serovars can be important given differences in antibiotic therapy compared to non-LGV disease.

Tests

- Multiple automated and semi-automated qualitative detection duplex tests using a variety of amplification chemistries (real-time PCR, strand displacement amplification, transcription-mediated amplification) have been cleared or approved.
- *C. trachomatis* assays detect all serovars but do not distinguish serovars D-K from LGV serovars.
- Specific tests for LGV serovars are laboratory-developed qualitative detection tests.

Sample Types

Samples for use in US FDA-cleared/approved tests vary by assay; package inserts should be consulted for specific assays. Specimens generally tested include:

- Males: First void urine (urethra also approved, but second-line due to ease of collection and equivalent sensitivity of urine).
- Females: First void urine, urethral swabs, cervical swabs, self-collected vaginal swabs, liquid-based cervical cytology medium.
    - Self-collected vaginal swabs have the highest sensitivity for detecting *C. trachomatis* infection in a given patient; thought to be due to vaginal pooling from cervical and urethral sites.
- Rectal, anal, and throat swabs are useful for detection of infection at these sites, but require validation by individual laboratories as they have not been approved for use in commercial assays.
    - These specimens are particularly useful for diagnosing LGV.

Limitations

- *C. trachomatis* strains with cryptic plasmid deletions are circulating in Scandinavia and Europe.
- False-negative results can be obtained with assays amplifying only cryptic plasmid.

Herpes Simplex Viruses

HSV-1 and -2 are DNA viruses that are transmitted by contact with infected skin or mucosa. After acute (primary) infection, viruses establish latent infection in sensory ganglia (sacral ganglia in genital herpes, trigeminal ganglia in orolabial herpes). Reactivation of replication results in recurrent disease. Primary infection and recurrences are often asymptomatic or mild and are therefore commonly unrecognized. After asymptomatic primary infection and decades of quiescence, reactivation can occur leading to misperception of newly acquired primary infection.

Symptoms

- Classically, genital, perianal, or oral blisters that rupture and develop into painful ulcers that heal over 2–3 weeks.
- Symptoms during primary infection typically are more severe than recurrent disease.
  - Greater number of lesions over extensive area; can be accompanied by meningitis symptoms (fever, headache, stiff neck).

Diagnosis

- Conventional methods are still standard in most laboratories.
- Nucleic acid detection tests are more sensitive than conventional methods, particularly when low levels of virus are present as in recurrent disease or healing lesions.
  - Molecular tests are poised to become the diagnostic standard given enhanced sensitivity compared to culture and number of

assays under development for use with automated "sample in/ answer out" instruments.

Treatment

- Both viruses are treated with oral acyclovir or valacyclovir.
- Treatment indications:
  - During acute symptomatic episodes: To shorten symptom duration in primary and recurrent disease.
  - Daily suppressive therapy to prevent:
    - Reactivation in individuals with a history of severe recurrent disease.
    - Transmission (along with barrier contraception) to uninfected partner.

HSV-1 vs HSV-2

- HSV-1 and HSV-2 replicate most efficiently in orolabial and genital mucosa, respectively; however, both viruses can cause disease at both sites.
- Genital HSV-1 infections are typically less severe than HSV-2:
  - Less severe symptoms
  - Fewer recurrences
- Differentiation between the two viruses is diagnostically necessary in genital herpes to counsel patients regarding:
  - Expected disease severity
  - Need for future suppressive therapy (more common with HSV-2 than HSV-1)

Tests

- US FDA-cleared qualitative multiplexed nucleic acid detection tests with varying chemistries (real-time PCR, strand displacement amplification) are available.

Sample Types

- Accepted specimens vary by commercial assay; package inserts should be consulted.
- Generally, anogenital lesion swabs submitted in viral transport medium are useful; vaginal swabs are accepted for certain assays.

Limitations

- Tests that do not distinguish HSV-1 from HSV-2 have impaired diagnostic utility (see HSV-1 vs HSV-2 above).
- VZV reactivation (herpes zoster) can manifest with symptoms similar to genital infections with HSVs and should be considered in individuals with lesions and no detectable HSV-1 or -2 by nucleic acid amplification test.
- Qualitative detection of HSV-1 or -2 cannot distinguish between primary and recurrent infection.

Trichomonas vaginalis

T. vaginalis is a flagellated parasitic protozoan that causes genital infections that are often asymptomatic but can manifest in women as vulvar or vaginal inflammation and in men as nongonococcal urethritis. The diagnosis should trigger partner investigation and treatment to avoid reinfection that occurs commonly in the absence of intervention. Complications in pregnant women include premature rupture of membranes and pre-term delivery.

Symptoms

- Women: Vulvar/vaginal itching, burning, dyspareunia; inflamed mucosa, discharge (clear, white, yellow or green, often malodorous and copious).
- Men: Burning on urination; penile discharge.

Treatment

Oral nitroimidazole antibiotics (metronidazole, tinidazole).

Tests

Qualitative nucleic acid detection is indicated and is more sensitive than conventional methods. US FDA-cleared formats include:

- Direct hybridization (as part of a triplex assay in combination with *Gardnerella vaginalis* and *Candida albicans* for use in patients with vaginitis).
- Nucleic acid amplification (transcription mediated amplification), which has been reported to be more sensitive than direct hybridization.

Sample Types

Genital samples including vaginal swabs, endocervical swabs, endocervical samples in liquid-based cervical cytology medium, and urine. Package inserts should be consulted for specific tests.

Limitations

Screening asymptomatic adolescents, adults and pregnant women is not yet recommended despite the prevalence of asymptomatic infection.

Human Papillomavirus

High-Risk Types

HPVs are double-stranded DNA viruses that are classified into more than 100 types based on biologic and genetic characteristics. They infect cutaneous and mucosal epithelium. The majority of infections are subclinical and self-resolving. Low-risk HPV genotypes (LR HPV) cause benign lesions, primarily warts; fourteen high-risk genotypes (HR HPV) are associated with cancer (Table 13). Anogenital disease is largely transmitted by sexual contact although intercourse is not

Table 13. US FDA-Approved Tests for High-Risk HPV Detection

Test (Manufacturer)	*Detection Chemistry*	*HPV Types Detected*	*Comment*
Aptima HPV (Hologic Gen-Probe)	Transcription mediated amplification	16, 18, 31, 33, 35, 39, 45, 51, 52, 56, 58, 59, 66, 68	HPV E6/E7 messenger RNAs are amplified, which may enhance clinical specificity in ASC-US. US FDA-approved genotype assay is available for this platform.
Cervista HPV HR (Hologic)	Invader (target amplification)	16, 18, 31, 33, 35, 39, 45, 51, 52, 56, 58, 59, 66, 68	US FDA-approved genotype assay is available for this platform.
Hybrid Capture 2 (hc2, Qiagen)	Hybridization (signal amplification)	16, 18, 31, 33, 35, 39, 45, 51, 52, 56, 58, 59, 68	No US FDA-approved genotype assay available for this platform.
Cobas 4800 HPV (Roche Molecular Diagnostics)	Real-time PCR	16, 18, 31, 33, 35, 39, 45, 51, 52, 56, 58, 59, 66, 68	Multiplex format; detects HR HPVs and individually identifies types 16 and 18 in a single assay.

ASC-US, atypical squamous cells of undetermined significance; FDA, Food and Drug Administration; HPV, human papilloma virus; HR, high risk.

required. LR HPV types (primarily types 6 and 11) cause benign anogenital warts and low-grade squamous intraepithelial lesions (LSILs) in cervical Papanicolaou stains (Pap smears). Anogenital infections with HR HPV types are also predominantly subclinical, self-resolving, and cause LSIL. A small proportion of these infections can persist and cause precancerous lesions (termed high-grade intraepithelial lesion [HSIL] by Pap smear cytology; on biopsy, termed cervical intraepithelial neoplasia [CIN] grade 2 or the higher grade 3) that can evolve into invasive cancer. Tumors can occur at any genital site; the most common is squamous cell carcinoma in the squamocolumnar junction of the cervix, followed by adenocarcinoma and adenosquamous cell carcinoma.

Given the association between HR HPV and cervical cancer, detection of HR HPV is recommended in management guidelines in the following instances:

- In initial triage management of women ≥ 21 y of age with atypical squamous cells of undetermined significance (ASC-US; use of HR HPV testing in this setting is termed ASC-US with reflex HR HPV).
  - The major utility of reflex HR HPV in ASC-US is the high negative predictive value (NPV) for underlying cervical cancer when no HR HPV is detected (NPV, 98%).
  - The detection of HR HPV in specimen with ASC-US cytology is not necessarily indicative of underlying cervical cancer (positive predictive value ~65%). These women should be managed as per American Society for Colposcopy and Cervical Pathology (ASCCP) guidelines.
- For women ≥ 30 y of age, in conjunction with Pap smear (termed cotesting).
  - Women with normal cytology and no detectable HR HPV can defer Pap smear for 5 y due to extremely low rates of cervical cancer in this cohort over this interval.
  - Women with normal cytology and detectable HR HPV should have repeat cotesting in 12 months or reflexive HR HPV genotype testing.

    - Women with HPV 16 or 18 should be referred directly for colposcopy due to high risk of underlying cervical cancer which was not detected cytologically most likely due to sampling error (see *Genotyping Assay* below for a description of these tests).

- For postmenopausal women with LSIL cytology; currently any of the following is recommended in this special population:
    - Reflex HR HPV testing
        - If no detectable HR HPV → repeat cytology in 12 months
        - If HR HPV detectable → colposcopy
    - Repeat cytology at 6 and 12 months
    - Colposcopy
- AFTER colposcopy in the following settings:
    - Women of any age with initial cytology of atypical glandular cells (AGCs) or atypical squamous cells: Cannot rule out HSIL (ASC-H) when no evidence of cervical intraepithelial neoplasia (CIN 2,3) found at colposcopy (due to unsatisfactory colposcopy, no lesion identified at colposcopy, or lesion identified, but biopsy did not show CIN2 or 3).
    - Women ≥21 y of age with initial ASC-US or LSIL cytology when no evidence of CIN 2,3 found at colposcopy (due to unsatisfactory colposcopy, no lesion identified at colposcopy, or lesion identified, but biopsy did not show CIN2 or 3).
    - In the above two settings, postcolposcopy management options include:
        - HR HPV testing 12 months after colposcopy; or
        - Repeat cytology 6 and 12 months after colposcopy.

Tests

Multiple platforms with varying qualitative detection chemistries have been approved for use by the US FDA (Table 13). See *Genotyping Assays for the Management of Infectious Diseases* for additional information on HR HPV genotyping tests.

Sample Types

Proprietary collection media and liquid-based cervical cytology medium (PreservCyt) have been approved for use in these assays.

Limitations

- HR HPV testing **should not be performed and should not be used for management** in the following settings:
  - To guide HR HPV vaccination decisions.
  - In initial management or triage of adolescents (girls ≤ 20 y of age) with any abnormal cytologic result.
  - In initial triage of LSIL (except postmenopausal women, see above), ASC-H, HSIL, AGC, or adenocarcinoma *in situ*.
- There is currently no role for LR HPV testing in cervical cancer screening.
- False-positive results have been observed with the hybridization-based test (hc2, Qiagen) due to detection of some low risk types.
- HR HPVs are associated with anal cancers, however HR HPV detection tests are not approved for anal cancer screening or diagnosis.

Health Care–Associated Infections

C. difficile

C. difficile infection (CDI): Disease caused by toxins A and B (TcdA and TcdB) encoded by bacterial *tcdA* and *tcdB* genes.

- Animal studies suggest TcdB is more important for CDI than TcdA.
  - TcdB-/A+ and TcdB+/A- strains are uncommon but have been found in CDI.
- Endemic and epidemic infections occur. Endemic CDI is found in high-risk populations:
  - Antibiotics within past 3 months

  - Hospitalization/long-term care facility (can present many weeks after discharge)
  - Advanced age
  - Chemotherapy
  - Gastrointestinal (GI) surgery, tube feeding, autoimmune colitis
  - Gastric acid reduction therapy (proton pump inhibitors) is an emerging risk factor
- Epidemic CDI: Community-acquired disease in the absence of risk factors.
  - Due to highly virulent strain (referred to as NAP1 or Ribotype 027)
  - Occurs in populations at low risk of endemic CDI (young individuals, pregnant women)
  - Uncommon compared with endemic CDI

Symptoms

- Mild to moderate CDI: fever, watery diarrhea, abdominal pain.
- Severe CDI: above symptoms plus white blood cell count > 15,000/mm^3 or $\geq$ 50% increase in creatinine.
- Severe, complicated CDI: severe disease with hypotension, ileus, toxic megacolon, colon perforation, colectomy, admission to the intensive care unit (ICU).

Treatment

- Oral metronidazole for mild/moderate CDI.
- Oral vancomycin for severe CDI.
- Oral vancomycin plus IV metronidazole for severe complicated CDI.

Recurrent Infection

Approximately 20% relapse rate one week to two months after standard therapy.

- Due to reinfection with new strain or recrudescence of initial infection.
  - Current antibiotics kill cells, not spores; spores can survive up to five months and can reinitiate infection.
  - Recurrence due to antibiotic resistance is uncommon but can occur with metronidazole.
- Clinical symptoms are the same as initial infection.
- A positive test result should be obtained prior to retreatment.

Tests

Qualitative detection *tcdB* and/or *tcdA* (depending upon the assay). Chemistries include real-time PCR and loop-mediated isothermal amplification.

Sample Types

Semi-solid/liquid stool; testing on formed stool is contra-indicated.

Limitations

- Testing in asymptomatic individuals is not useful, including for test-of-cure.
- False-negative NAAT results can occur in CDI.
  - Sensitivity of NAAT is 90%–97% compared with the gold standard for sensitivity (toxigenic culture, anaerobic culture plus toxin detection by cell culture cytotoxicity assay).
  - *tcdB* deletion mutations (TcdB-/A+ strains) can produce false-negative results in *tcdB*-based assays; *tcdA* deletion mutations should still be detected by a *tcdA*-based assay; the target sequence lies outside the deletion domain.
  - Consider toxigenic culture if NAAT is negative and clinical suspicion is high.
- False-positive results can occur due to asymptomatic colonization with toxin-producing strains.

- Positive cell cytotoxicity rates of < 8%, < 5%, and ~60% have been found in hospitalized patients, asymptomatic antibiotic-treated adults, and healthy neonates with positive *C. difficile cultures*.
- Testing should be restricted to symptomatic individuals to assure accurate result interpretation.

Methicillin-Resistant Staphylococcus aureus

β-Lactam antibiotics (penicillins and cephalosporins) bind and inhibit transpeptidases, enzymes that function in bacterial cell wall synthesis. *mecA* encodes a transpeptidase (penicillin binding protein 2a [PBP 2a]) that has reduced affinity for β-lactam antibiotics. Cell wall synthesis is not impacted in *mecA*-expressing strains. Methicillin resistance is chromosomally encoded by *mecA* in the staphylococcal chromosomal cassette (SCC*mec*) mobile genetic element.

Test Utility

- Prevention of healthcare-associated methicillin-resistant *S. aureus* (MRSA):
  - Major use is to identify colonized individuals at or before hospital admission as part of a comprehensive infection control plan to prevent nosocomial MRSA infections due either to transmission or autoinfection.
  - Favorable cost/benefit outcomes depend on regional/institutional MRSA prevalence and decolonization rates.
  - Comprehensive screening of all admissions has been adopted voluntarily on a limited basis and where legislated.
  - Guidelines advocate targeted screening of patients at risk of nosocomial MRSA complications where MRSA prevalence and complication rates are high.
    - Elective surgery (including cardiac surgery) patients to prevent surgical site wound infections.
    - ICU patients to prevent bloodstream infections due to indwelling catheters.

- Intranasal mupirocin and chlorhexidine baths can be instituted for decolonization therapy to help decrease nosocomial MRSA.
- Diagnosis of MRSA infections in symptomatic patients:
  - Rapid, qualitative detection of MRSA nucleic acid eliminates the need for empiric antibiotic therapy for MRSA coverage while awaiting conventional susceptibility test results.
    - Cost/benefit data for rapid testing are lacking.

Tests

Qualitative nucleic acid detection is performed. Tests utilizing a variety of nucleic acid detection chemistries are available including target amplification (real-time PCR and nucleic acid sequence-based amplification) and direct detection through hybridization (with signal amplification). *mecA* target strategies vary by assay and include:

- Amplification of the junction between SCC*mec* cassette and its site of insertion in the *S. aureus* chromosome (*ORF-X)* as an indirect indicator of *mecA*.
- Dual target approach: *ORF-X*/SCC*mec* junction and *mecA* sequence amplification.
- Triple target approach: *S. aureus* amplification (*S. aureus*-specific gene target) and *ORF-X*/SCC*mec* junction and *mecA* sequence amplification.

Separate tests are available for MRSA surveillance and for disease diagnosis (skin/soft tissue abscess, sepsis).

Sample Types

- MRSA surveillance tests: Nares swabs; MRSA is found in nares of ~75% colonized individuals.
- MRSA diagnostic tests:
  - Skin/soft tissue infections: Swabs of purulent material from infected site.
  - Sepsis: Blood cultures known to contain Gram-positive cocci.

Limitations

- False-negative results can be observed in individuals:
  - Colonized at sites other than nares; ~25% of colonized individual have transmissible MRSA at other sites (skin, throat).
  - With variant SCC*mec* sequences; five SCC*mec* types (I–V) exist, with variable genetic constituents in addition to *mecA*. An inability of some assays to detect type V has been reported.
- False-positive results are observed when *mecA* is deleted from SCC*mec* cassette.
  - SCC*mec* extremity and chromosomal sequences adjacent to SCC*mec* integration site are retained in these mutants.
  - ~5% false-positive rate is observed for assays using this target strategy.

Vancomycin-Resistant Enterococci

Vancomycin binds cell wall peptidoglycan precursors preventing transpeptidation and transglycosylation required for cell wall function. Vancomycin resistance is mediated by multiple proteins (VanA-VanE, VanG); VanA and VanB are most common. VanA and VanB are encoded on transposons (plasmids) by separate multigene operons; resistance proteins alter peptidoglycan precursors resulting in decreased drug binding.

Sites of VRE infection are:

- Skin/soft tissue
- Blood stream (sepsis)
- Heart valves (endocarditis)
- Abdomen
- Urinary tract

Susceptible hosts include immunocompromised individuals, especially stem cell and solid organ transplant patients.

Test Utility

- Surveillance:
  - Used as rapid method for identification of colonized individuals at hospital admission as part of a comprehensive infection control plan to prevent nosocomial VRE infections.
  - Primary source of transmitted vancomycin-resistant enterococci (VRE) is colonized patients via carriage to other patients on health care workers' hands and clothing.
    - Transmission from contaminated environment can also occur.
  - Transmission prevention methods include identification/isolation (with contact precautions) of colonized individuals.
  - Guidelines advocate targeted screening of patients at risk of VRE colonization (individuals with prior exposure to acute or chronic care facilities).
  - VRE NAAT use for screening is not yet widespread; most surveillance performed by culture.
    - Rapid time to result compared to culture could permit faster implementation of isolation practices.
- Diagnosis of sepsis

Tests

Qualitative nucleic acid detection is performed. Real-time PCR is available for surveillance; direct detection of *vanA* and *vanB* through hybridization (with signal amplification) is available for sepsis diagnosis.

Sample Types

- Surveillance: Perirectal or rectal swabs.
- Sepsis diagnosis: Blood cultures known to contain Gram-positive cocci.

Limitations

- False negatives: Interpretive comment regarding the potential for colonization with a vancomycin-resistant organism expressing VanB should be included with results of assays detecting *vanA* only.
- False positives: Intestinal anaerobes can express VanB and may yield false-positive VRE results in assays that detect *vanB*. Clinical implications of such false positives are moderated by fact that transposon-encoded *vanB* operon may be readily transmitted to co-residing enterococci.

Infections Managed Through Nucleic Acid Testing of Blood

Nucleic acid testing of blood is currently the standard of practice for the diagnosis and management of many viral infections; testing for other types of pathogens is relatively limited. Assays to detect and quantify viral nucleic acids in blood are useful for infections in which viremia serves as a noninvasive marker of viral replication when a single organ is infected and for others in which it is an indicator of dissemination that results in multiple diseased organs. This section focuses on the most common viral infections that are managed by molecular testing of blood. Background information (viruses, susceptible hosts/diseases, treatment, test utility) necessary to fully understand molecular testing is reviewed in addition to types of molecular tests, sample types, and test limitations.

Adenovirus

- **Virus:** Nonenveloped virus with double-stranded DNA genome. Different tissues are infected depending upon virus serotype.
- **Susceptible hosts/diseases:** Patients with numeric or functional T-cell deficits, including hematopoietic stem cell transplant (HSCT)

patients and solid organ transplant patients. Diseases seen in both populations: Fever of unknown origin, pneumonia, hepatitis, cystitis, colitis, encephalitis (rare).

Treatment

Ribavirin, cidofovir.

Test Utility

- Nucleic acid quantification:
  - Detection of disseminated infection in setting of fever of unknown origin, pneumonia, hepatitis, colitis
  - Therapeutic monitoring
- Qualitative nucleic acid detection:
  - Diagnosis of hemorrhagic cystitis, colitis, encephalitis

Tests

Current assays for nonrespiratory specimens are laboratory-developed real-time PCR tests for qualitative detection and quantification. No US FDA-approved tests are available (excluding respiratory virus assays).

Sample Types

- Quantitative tests: Plasma, whole blood.
- Qualitative tests: Urine (hemorrhagic cystitis), stool (colitis), CSF (encephalitis).

Limitations

- No international standard is available as a calibrator for quantification; measured levels could be different with different assays. Monitoring of individual patients should therefore be performed with a single assay.

- In the absence of an international standard calibrator, globally applicable quantitative thresholds correlating with disease in different patient populations have not been determined.
- Qualitative testing of whole blood has limited diagnostic utility given adenovirus latency in lymphocytes.
    - Quantification is required to distinguish latency from active replication.
        - Cutoffs that distinguish between latent infection and active replication in whole blood samples should be established by individual laboratories.
        - No universally applicable cutoff is available.

BK Virus

- **Virus:** Nonenveloped virus with double-stranded DNA genome. Mode of transmission is unknown; likely respiratory. During acute infection, it infects renal tubular cells where it remains latent. Reactivation of replication occurs during immunosuppression.
- **Susceptible hosts/diseases:**
    - Nephropathy (infection and destruction of renal tubular cells) is seen in ~5% of renal transplant patients, mostly (~75%) in first-year post-transplant. Prior immunity is not protective; nephropathy is observed commonly in BK-seropositive renal allograft recipients. Reported nephropathy-associated allograft loss rates are $\geq$ 10%.
    - Hemorrhagic cystitis: urinary frequency, urgency, dysuria, with varying degrees of hematuria. Seen in 10%–25% HSCT patients.

Treatment

- Nephropathy: Reduction of immunosuppression.
- Hemorrhagic cystitis: Supportive care including symptom relief, hydration, and bladder irrigation.

Test Utility

- Quantification, used primarily in nephropathy:
  - Detection of viral replication allows for intervention (reduction in immunosuppression, administration of cidofovir) to prevent BK virus–induced nephropathy and allograft loss in renal transplant patients.
    - Biopsy is recommended for BKV DNA > 7.0 $\log_{10}$ copies (urine) or > 4.0 $\log_{10}$ copies (plasma).
      - After therapeutic intervention, further monitoring at 2–4-week intervals is recommended to document decline in viremia.
- Qualitative detection: Used primarily for the diagnosis of BK virus–induced hemorrhagic cystitis.

Tests

Nucleic acid quantification and qualitative detection. No commercially available tests. Current assays are largely laboratory-developed real-time PCR tests.

Sample Types

- Quantitative tests: Plasma, urine.
- Qualitative tests: Urine.

Limitations

- No international standard is available as a calibrator for quantification; measured levels could be different with different assays. Monitoring should therefore be performed with a single assay.
- In the absence of an international standard calibrator, quantitative threshold that triggers renal biopsy should be established by individual laboratory.
- False negatives can occur.

- BK virus strains are considerably more genetically diverse than originally appreciated.
- Assays should be carefully validated analytically to ensure efficient amplification/quantification of all genotypes.

Cytomegalovirus

- **Virus:** Enveloped virion with double-stranded DNA genome; member of *Herpesviridae*.
  - Acute infection occurs commonly during childhood by contact with infected secretions and is largely asymptomatic. Infected cells include epithelial cells, endothelial cells, peripheral blood mononuclear cells, and neurons. Latency is established in peripheral blood mononuclear cells after clearance of acute infection.
  - Infection can also be transmitted transplacentally, mostly during acute (primary) maternal infection.
  - Replication reactivation occurs during immunosuppression. End-organ disease produced by reactivation varies by host (see below) but can include encephalitis/retinitis (see *Central Nervous System Infections* for additional diagnostic information), pneumonitis, hepatitis, GI disease (anywhere from esophagus to colon).
- **Susceptible hosts/diseases:**
  - HIV: End-organ disease, particularly encephalitis/retinitis/GI disease in advanced disease (CD4<50 cells/mm^3). Pneumonitis does not occur, although CMV is readily detectable in the lower respiratory tract.
  - HSCT: Any end-organ disease, usually after engraftment (30–60 days post-transplant); now prevented through prophylaxis with valacyclovir. CMV seropositive HSCT with T-cell deficits are at highest risk.
  - Solid organ transplant recipient: "CMV syndrome" (fever, malaise, leucopenia, thrombocytopenia) or end-organ disease (particularly of the allograft in lung, liver, bowel transplant

patients). CMV seronegative recipients of allograft from CMV seropositive donor (D+/R–) are at highest risk. Disease prevention strategies in D+/R–:

- Prophylaxis with valganciclovir: Duration dictated by disease risk; determining factors include transplant type and CMV serostatus. High-risk patients: D+/R– lung transplant à 12 months; D+/R– renal transplant → 6 months. **Viral load monitoring after prophylaxis is critical as late-onset disease observed in ~20%.**
- Preemptive therapy: Monitoring for CMV replication at regular intervals post-transplant with initiation of valganciclovir after viremia detection.

- Fetus (congenital infection):
  - Majority (95%) are overtly asymptomatic at birth.
  - Sensorineural hearing loss is most common (~20% asymptomatic and one-third of symptomatic babies).
  - Severe symptoms more commonly observed in congenital infections acquired from mothers with primary infection (vs reactivation). Symptoms include intrauterine growth retardation, hepatosplenomegaly, petechiae, jaundice, CNS complications (microcephaly, seizures, calcifications on computed tomography [CT] scan), ocular abnormalities (chorioretinitis, retinal scars, optic atrophy, blindness).
- Immunocompetent adolescents and adults: infectious mononucleosis syndrome. Symptoms include fever, fatigue, malaise, adenopathy, pharyngitis (less common than in EBV mononucleosis). Laboratory testing demonstrates:
  - Lymphocyte abnormalities (lymphocytosis or lymphopenia).
  - Hepatitis (elevated alanine aminotransferase, aspartate aminotransferase).
  - No serologic evidence of EBV infection (heterophil "Monospot" antibody negative, no detectable immunoglobulin [Ig] M/IgG to EBV viral capsid antigen).

Treatment

Usually only in immunocompromised individuals.

- First-line: Ganciclovir (or its orally bioavailable for valganciclovir).
- Second-line (after demonstration of ganciclovir resistance): Foscarnet, cidofovir.

Test Utility

- **Nucleic acid quantification** (peripheral blood specimens):
  - Diagnosis of end organ disease in immunocompromised patients, as surrogate for specimens collected with invasive procedures (primarily biopsy).
  - To detect and treat viral replication prior to disease onset in solid organ transplant recipients (disease prevention through preemptive treatment).
  - Viral load monitoring to assess therapeutic response during treatment.
  - Resistance detection: Continued or resurgent replication during treatment. Should be confirmed by definitive genotype assay (see *Genotype Testing*) to guide management with second-line agents.
- **Qualitative nucleic detection** (specimens other than peripheral blood):
  - Detection of congenital infection *in utero* from amniotic fluid.
  - Detection of CNS (see *Central Nervous System Infections*) or ocular infection.

Tests

- **Nucleic acid quantification:** currently one US FDA-approved commercial assay available. Laboratory-developed tests are common as quantitative testing in immunocompromised patients is the standard of care.

- The first World Health Organization international standard for CMV is available for use in calibration. Standard preparation is freeze-dried cell-free virus (prototype Merlin strain) more suitable for use in calibration of plasma-based rather than cell-based (whole blood) quantitative assays.
- Standardized calibration and reporting of quantitative data in IU/mL should facilitate quantitative uniformity across laboratories and allow for the development of broadly applicable cutoffs for use in patient management.

- **Qualitative nucleic acid detection:** No commercially available tests. Current assays are largely laboratory developed.

Sample Types

- **Quantitative tests:** Peripheral blood samples (plasma, whole blood).
    - Viral loads in whole blood are higher and CMV viremia is detectable earlier in whole blood than plasma.
    - During treatment, clearance of CMV DNA from whole blood is slower compared with plasma, yet positive predictive values for recurrence are similar.
        - Monitoring therapeutic response in whole blood may therefore lead to unnecessarily prolonged treatment.
    - Overall, there appears to be no clear clinical advantage to whole blood over plasma and plasma is used extensively for patient management.
    - Given quantitative differences between whole blood and plasma, individual patients should be followed with a single sample type.
- **Qualitative nucleic acid detection tests:**
    - Amniotic fluid, urine (from affected newborn) for suspected congenital infection.
    - Tissue can be tested for end-organ disease but is not widely available.

Limitations

- Quantitative thresholds in IU/mL for guiding management decisions in immunocompromised individuals are not yet available despite the availability of an international quantification standard.
  - Data for quantitative guidelines are lacking; studies incorporating quantification in IU/mL in the management of CMV infections have not yet been extensively performed.
- Qualitative testing of whole blood has limited diagnostic utility given CMV latency in peripheral blood mononuclear cells.
  - Quantification is required to distinguish latency from active replication. Cutoffs that distinguish between latent infection and active replication in whole blood samples should be established by individual laboratories. No universally applicable cutoff is available.
- False-negative quantitative results from blood can be observed in end-organ disease, particularly in GI disease.
  - For GI presentations, biopsy should be considered if no CMV DNA is detected in peripheral blood and clinical suspicion remains high. Empiric antiviral therapy may be an option if biopsy cannot be performed.
  - Quantitative assessment of viremia lacks adequate clinical sensitivity for CNS and ocular disease. CSF and vitreous fluid specimens are preferred specimens for these presentations.
- CMV DNA concentrations vary by peripheral blood compartment (CMV DNA levels in cellular samples are typically greater than plasma).
  - CMV DNA quantification should therefore be performed in a single sample type to facilitate interpretation of quantitative trends in individual patients.
- Quantitative testing of bronchoalveolar lavage and tissue has been advocated by a recent solid organ transplantation guideline, primarily due to increased sensitivity and faster time to result than culture.

  - Lack of quantitative interpretive cutoffs (as acknowledged by the Transplantation Society International CMV Consensus Group and the guidelines on the management of CMV in solid organ transplantation) lessens the current overall utility of these assessments.
- Quantitative testing of blood can be useful in diagnosis of uncomplicated mononucleosis if testing is performed early after symptom onset.
  - False-negative results can be observed if testing is delayed due to clearance of viremia.
- Quantitative testing of blood from infants is not commonly used in the United States for the management of congenital infections.

Epstein Barr Virus

- **Virus:** Enveloped virion with double-stranded DNA genome; member of *Herpesviridae*.
  - Acute infection occurs commonly during childhood by contact with infected secretions and is largely asymptomatic.
  - Infected cells include B lymphocytes, epithelial cells, and monocytes.
  - Latency is established in B lymphocytes after clearance of acute infection.
- **Susceptible hosts/diseases:**
  - Immunocompetent children, adolescents, adults: Infectious mononucleosis (fever, fatigue, malaise, pharyngitis, adenopathy, atypical lymphocytosis) due to acute infection. Children are mostly asymptomatic but can have mononucleosis. Serology (heterophil "Monospot" antibody) is the diagnostic gold standard. Children < 2 y of age can be difficult to diagnose with serology; heterophil antibody is typically negative in most cases and viral capsid antigen (VCA) IgM detectable in ~60%.
    - Rare complications of infectious mononucleosis

- Fatal infectious mononucleosis (FIM): Proliferation of EBV-infected B-cells, cytotoxic T cells, macrophages with tissue invasion (liver, bone marrow, heart, brain), abnormal cytokine release; results in tissue damage and organ dysfunction. Seen in males with X-linked lymphoproliferative disease (XLPD); ~60% of XLPD due to mutations in SH2D1A gene encoding SLAM-associated protein, a signal transduction molecule.
- Hemophagocytic lymphohistiocytosis (HLH): Abnormal proliferation of T cells and histiocytes after acute viral infection. EBV is a common inciting infection but other viruses are also implicated. Occurs due to mutations impairing perforin-mediated T-cell (CD8)-mediated cell killing. CD8 cell cytotoxicity clears infected cells and downregulates antigen-presenting, cytokine-producing T cells and histiocytes. Unchecked, these cells proliferate and infiltrate tissues, producing cytokine storm. Histiocyte ingestion of red cells, white cells, and platelets in bone marrow, lymph nodes, and spleen are histologic hallmarks.

- Depending on host immune status, EBV is associated with a variety of tumors.
  - Immunocompromised patients:
    - Solid organ transplant, HSCT (T-cell functional or numerically depleted hosts): Post-transplant lymphoproliferative disease, a heterogenous group of lymphoid disorders ranging from infectious mononucleosis to monomorphic lymphoid tumors.
      - Highest risk associated with primary EBV infection (therefore, most commonly observed in pediatric patients).
    - AIDS: B-cell lymphomas, oral hairy leukoplakia (EBV replication in lingual squamous epithelial cells causing painless white patches on lateral aspects of tongue).

- Immunocompetent individuals: Lymphoid tumors (endemic Burkitt lymphoma, ~30% of Hodgkin disease) and carcinoma (nasopharyngeal carcinoma [NPC]).

Treatment

- Infectious mononucleosis (IM): Supportive care. Rituximab (anti-CD20 monoclonal antibody) has efficacy for EBV-associated HLH and in acute EBV infections in individuals XLPD.
- Lymphoma: Conventional oncologic management. Antiviral therapy not useful.
- NPC: Radiation, chemotherapy.
- PTLD: Prophylaxis with antiviral drugs (high-risk recipients, immediately post transplant prior to EBV detection in peripheral blood), immunosuppression reduction or anti-CD20 antibody (rituximab) to prevent disease after EBV DNA detection in blood; rituximab or other standard chemotherapy regimens for disease treatment.

Test Utility

Quantitative testing for EBV DNA in blood can be helpful in the following settings:

- IM: Immunocompetent adolescents/adults early in acute infection prior to antibody production; infants, children < 2 y of age, immunocompromised individuals with impaired serologic responses.
  - FIM/HLH: EBV DNA measurement in peripheral blood in patients with consistent clinical presentations confirms disease etiology. These diseases are associated with extremely high EBV DNA levels.
- Post-transplant lymphoproliferative disorder (PTLD): The use of post-transplant monitoring for EBV in blood of high-risk patients for PTLD prevention and early detection has been recognized by recent guidelines. Testing has been extended to therapeutic response assessment and disease recurrence detection.

- Lymphoma: Nonmolecular assays are the standard for management of non-Hodgkin lymphoma and Hodgkin's disease however EBV viral load is being used increasingly as adjunctive marker of disease activity, response to therapy, recurrence.
- NPC: Literature meta-analysis suggests utility as noninvasive marker for this cancer, potentially useful as screening tool in high prevalence areas (primarily Asia).

Tests

- Quantitative nucleic acid tests are typically used.
  - The first World Health Organization international standard for EBV is available for use in calibration.
  - Standard preparation is freeze-dried cell-free virus (prototype B95-8 strain).
    - International standard may be more suitable for calibration of plasma-based rather than cell-based (whole blood) quantitative assays.
    - Standardized calibration has not yet been widely adopted, despite the availability of this preparation.
- Qualitative tests can be difficult to interpret due to latent EBV infection in B lymphocytes.
- No US FDA-approved tests are available. Most assays in use are laboratory-developed real-time PCR tests.

Sample Types

The relevant peripheral blood sample type is disease-dependent, related to pathogenesis, specifically EBV DNA localization and state (Table 14).

Table 14. Localization of Epstein Barr Virus DNA

Disease	*Site of EBV DNA*	*EBV DNA State*	*Optimal Peripheral Blood Compartment*
IM	Replicating virus in acutely infected epithelial cells and lymphocytes	Circulating cell-free virions, viral DNA	Plasma
PTLD	B lymphocytes	Circulating B lymphocytes, cell-free DNA from involved apoptotic cells	PBMCs, plasma, or whole blood[a]
Lymphoma	Non-circulating tumor cells	Cell-free DNA from apoptotic tumor cells	Plasma
NPC	Non-circulating carcinoma cells	Cell-free DNA from apoptotic tumor cells	Plasma

[a]Optimal peripheral blood compartment for use in PTLD management is controversial. PBMCs were used in original studies establishing quantitative assessment utility. Available data suggest cellular specimens are sensitive but not specific for PTLD detection; plasma may be adequately sensitive and have optimal specificity. Many centers utilize whole blood as a diagnostic compromise and to simplify workflow (eliminate need to prepare PBMC or process plasma). Available guidelines support the use of quantification in PTLD management but stop short of recommending a particular sample type.

EBV, Epstein Barr virus; IM, infectious mononucleosis; PBMC, peripheral blood mononuclear cell; PTLD, post-transplant lymphoproliferative disorder; NPC, nasopharyngeal carcinoma.

Limitations

- Quantitative thresholds in IU/mL for guiding management decisions are not yet available despite the availability of an international quantification standard.
  - Data for quantitative guidelines are lacking; studies incorporating quantification in IU/mL in the management of EBV-associated disease have not yet been extensively performed.
  - Quantification is required to distinguish latency from active replication.
    - Cutoffs that distinguish between latent infection and active replication in whole blood samples should be established by individual laboratories.
    - No universally applicable cutoff is available.
  - EBV DNA concentrations vary by peripheral blood compartment (EBV DNA levels in cellular samples are typically greater than plasma).
    - EBV DNA quantification should therefore be performed in a single sample type to facilitate interpretation of quantitative trends in individual patients.
- Qualitative testing of whole blood has limited diagnostic utility given EBV latency in B lymphocytes.

Other Herpes Viruses (HSV, HHV Types 6 and 8, VZV)

Transmission/Infected Cells

- **HSV:** Common infection worldwide.
  - Transmitted through direct contact with infected skin or mucosa.
  - Virus infects neurons and epithelial cells; remains latent in sensory ganglia.
- **HHV-6:** Common infection worldwide.
  - Transmitted through contact with saliva during early childhood.

    - Virus infects T cells (primarily) and other cell types (monocytes, macrophage, epithelial cells, endothelial cells); remains latent in T cells.
- **HHV-8:** Endemic in Africa, with other geographic hotspots
    - In endemic regions, transmitted through contact with saliva and oropharyngeal secretions, usually during early childhood.
    - Sexual transmission (primarily among men who have sex with men) is most common route in low prevalence areas.
    - Primary sites of infection and latency are CD19+ B cells.
        - Other cells include endothelial-origin spindle cells, epithelial cells, macrophages.
- **VZV:** Common infection worldwide; now prevented through vaccination in childhood.
    - Primary infection transmitted via respiratory route.
    - Infects neurons, epithelial cells, endothelial cells.
        - Nerves within a number of different ganglia (cranial nerve, dorsal root, and autonomic) are sites of latency.

Susceptible Hosts/Diseases

- **HSV:**
    - Normal hosts: Primary and reactivation infections can be asymptomatic or manifest with skin lesions or mucosal ulcerations.
    - Immunocompromised hosts: Reactivation is more likely to be symptomatic; skin lesions, mucosal ulceration, disseminated disease with involvement of multiple organs, and encephalitis can occur.
- **HHV-6:**
    - Children: Infection is most commonly causes febrile rash illness (roseola infantum).
    - Immunocompromised hosts: Infection is most strongly associated with encephalitis and rash.

        - Other less common associations have been described (pneumonitis, hepatitis, colitis, CMV-like syndrome, bone marrow suppression/delayed engraftment).
- **HHV-8** causes tumors, including:
    - Kaposi sarcoma (caused by latent HHV-8 infection),
    - Castleman disease (plasmablastic variant, caused by active HHV-8 infection)
    - Primary effusion lymphoma (caused by latent HHV-8 infection).
- **VZV:**
    - Primary infection causes varicella ("chicken pox").
        - Pneumonitis, often fatal, is a rare complication.
    - Reactivation in the normal host
        - Herpes zoster (or "zoster"): Vesicular skin lesions in the dermatome innervated by the dorsal root ganglion where reactivation occurred is most common.
        - Reactivation from cranial nerve ganglia can cause eye and ear complications or facial paralysis.
    - Reactivation in immunocompromised hosts reactivation
        - Multidermatomal zoster
        - Disseminated disease with involvement of multiple organs and encephalitis
            - Can occur with or without skin lesions ("zoster sine herpete").

Treatment

- **HSV, VZV:**
    - Acyclovir (given prophylactically in hematopoietic stem cell transplant recipients due to high risk of reactivation/disease).
- **HHV-6:**
    - Ganciclovir or foscarnet; cidofovir is a second-line agent.

- **HHV-8:**
  - Kaposi sarcoma
    - Solid organ transplant recipients: reduction of immunosuppression.
    - HIV-infected patients: institution of antiretroviral therapy.
  - Castleman disease
    - Rituximab (anti-CD20 monoclonal antibody) with or without adjunctive chemotherapy.
    - In HIV patients, antiretroviral therapy should be initiated to prevent Kaposi sarcoma flares associated with rituximab use.
    - Addition of ganciclovir for suppression of HHV-8 replication also has efficacy.
  - Primary effusion lymphoma
    - Chemotherapy and institution of anti-retroviral therapy (in HIV patients).

Test Utility

- **HSV, HHV-6, and VZV:**
  - Detection in blood is useful for the diagnosis of disseminated disease and as a noninvasive surrogate for biopsy in immunocompromised patients.
- **HHV-8:**
  - Nucleic acid detection in blood is used to diagnose virus-associated malignancies.

Tests

- Quantitative nucleic acid tests are typically used.
- Qualitative assays of blood specimens have utility as initial tests for the detection of disseminated HSV and VZV since these viruses should be detectable in peripheral blood only when associated with disseminated or organ-based disease.

- No US FDA-approved tests are available.
    - Most assays are laboratory-developed real-time PCR tests.
    - HSV detection tests have been cleared for use in female genital specimens but not in blood or CSF.

Sample Types

Plasma and serum are useful for the detection of disseminated infections (HSV, HHV-6, and VZV) and HHV-8-related malignancies. Other sample types may be useful depending upon disease:

- Vesicle fluid, swabs from lesions (submitted in viral transport medium): HSV, VZV.
- CSF: HSV, HHV-6, VZV (see *Central Nervous System Infections* for further discussion).

Limitations

- International standards for use as calibrators to standardize quantification are not available for these herpesviruses.
    - Interassay variability in quantification is expected; patients should be monitored with a single assay.
    - Quantitative cutoffs to guide management are not available.
- Blood (and CSF) nucleic acid test results for HHV-6 can be difficult to interpret in the absence of accompanying chromosomal analysis tests such as fluorescence *in situ* hybridization (FISH) given viral genome integration in some individuals.
    - High viral loads (typically > 5.0 log_{10} copies/mL) are thought to be indicative of integrated virus.
    - Other causative pathogens should be definitively ruled out before invoking a diagnosis of HHV-6 when chromosomal analysis is not performed.
- Qualitative testing of whole blood has limited diagnostic utility given HHV-6 latency in T lymphocytes.

- Quantification is required to distinguish latency from active replication.
  - Cutoffs that distinguish between latent infection and active replication in whole blood samples should be established by individual laboratories.
  - No universally applicable cutoff is available.

Hepatitis B Virus

- **Virus:** Partially double-stranded DNA virus.
  - Multiple transmission routes
    - Blood-borne
    - Contact with secretions causes mother-to-child perinatal transmission and infection of young children by close contacts
    - Sexual transmission
  - Infects hepatocytes; hepatocyte destruction mediated by host immune response.
    - In chronic infection, HBV genome is retained in hepatocyte nucleus as covalently closed circular DNA.
- **Susceptible hosts/diseases:**
  - Immunocompetent individuals (young adult and older): Acute infection usually causes low-grade symptoms (fever, fatigue, abdominal pain), but can cause overt hepatitis (hepatomegaly and symptoms of hyperbilirubinemia including jaundice, acholic stools, dark urine); self-resolving in most (90%) cases.
  - Infants/young children, immunocompromised patients: Acute infection usually asymptomatic; most cases become chronic.
    - Recognized as an opportunistic infection in HIV. Incidence and rate of disease progression are greater in HIV/HBV coinfected than in HBV monoinfected individuals.
  - Four phases of chronic infection are summarized in Table 15.

Table 15. Four Phases of Chronic HBV Infection

Phase	*ALT/AST*	*HBsAg*	*HBeAg*	*Antibody to HBeAg*	*HBV DNA (IU/mL)*	*Liver Histology*	*Comment*
Immune tolerance	Usually normal	Present	Present	Absent	$\geq$ 20,000	Usually normal; can have mild inflammation	Usually the first phase of perinatal infection; can last for decades; patients usually asymptomatic.
Immune clearance	Elevated; can be episodic	Present	Present	Absent	$\geq$ 20,000	Active inflammation	Progression to cirrhosis and HCC related to disease severity during this phase.
Inactive HBsAg carrier	Usually normal; can have flares	Present	Absent	Present	< 2,000	Degree of abnormality dependent on disease' severity during clearance phase (mild inflammation to cirrhosis)	Outcomes include indefinite persistence, resolution of chronic infection (HBsAg clearance, appearance of anti-HBs), or disease reactivation due to re-emergence of original virus or HBeAg mutant virus (HBeAg- chronic hepatitis B).

HBeAg- chronic hepatitis B	Periodic flares	Present	Absent	Present	> or < 20,000	Active inflam- mation	Patients can progress directly from HBeAg+ to HBeAg- chronic hepatitis B. Mutant viruses have single base changes in core promoter and/ or precore region (stop codon).

Reprinted with permission from Valsamakis A. Molecular testing in the diagnosis and management of chronic hepatitis B. Clin Microbiol Rev 2007;20:426–39.

ALT, alanine aminotransferase. AST, aspartate aminotransferase. HBsAg, HBV surface antigen. HBeAg, HBV e antigen; HBV, hepatitis B virus; HCC, hepatocellular carcinoma.

Treatment

- Interferon alfa/pegylated interferon alfa: Mechanism of action is immune-mediated virus clearance including covalently closed circular DNA. A cure can therefore be achieved if patient is responsive. Patients with genotype A infections have higher response rates than other genotypes (see *Viral Genotyping Assays*).
- Nucleos/tide analogues (lamivudine, adefovir, telbivudine, tenofovir, entecavir currently approved for use): Mechanism of action is reverse transcriptase inhibition. Genome replication occurs via reverse transcription of full length RNA intermediate template. Reverse transcriptase inhibitors are therefore effective in suppressing replication but cannot cure infection since they do not eliminate covalently closed circular DNA (the template for full length RNA).
    - Tenofovir and entecavir are currently preferred due to high antiviral potency and low resistance rates.
    - In HIV/HBV coinfected patients, combination therapy (single pill) tenofovir + emtricitabine effective against both viruses is used commonly (although not approved).

Test Utility

- Discrimination between chronic, inactive hepatitis B and HBeAg-negative chronic hepatitis B.
- HBV DNA level in peripheral blood is one of several markers (including alanine aminotransferase [ALT], liver biopsy results) considered in deciding when to initiate treatment.
    - Risk of cirrhosis and hepatocellular carcinoma increases with increasing viral load.
    - Viremia thresholds for treatment initiation are variable among different guidelines.
    - Cirrhosis may be present in individuals with low viral loads therefore treatment initiation decisions should be individualized

and take into account multiple parameters (including ALT and liver biopsy).

- HBV DNA level is one of several markers (including HBeAg/HBeAb, HBsAg/HBsAb) used to monitor treatment response.
- During treatment, viral rebound after a period of undetectable viremia may indicate emergence of resistance and should trigger genotyping (see *Genotyping Assays*).

Tests

- Quantification methods such as real-time PCR are now preferred due to requirements for broad dynamic range (baseline viral loads can exceed 9.0 $\log_{10}$ IU/mL) and sensitivity (required to monitor treatment efficacy).
- Assays should be calibrated according to international standard and results reported as international units/mL.
- Several FDA-approved assays for use on automated platforms are available.
  - RealTime HBV (Abbott Molecular): 10 or 15 IU/mL depending on sample input volume to 9.0 $\log_{10}$ IU/mL.
  - Cobas AmpliPrep/Cobas TaqMan HBV (Roche Molecular Diagnostics): 20 IU/mL to 8.2 $\log_{10}$ IU/mL.

Sample Types

Plasma, serum.

Limitations

- Samples with viral loads greater than the approved upper limit of quantification require dilution; viral load result obtained after multiplication by dilution factor.
  - Occasional specimens remain above the upper quantification limit even after dilution and are therefore resulted as such.

- Specimens with HBV DNA levels greater than the upper quantification limit of NAAT pose contamination risks to the laboratory and to automated instrumentation.
    - Runs with samples in these high ranges should be scrutinized for evidence of contamination (occurrence of low positive results, typically ≤ 4.0 log10 IU/mL).
    - Samples with suspicious results should be re-tested to confirm original quantification.

Hepatitis C Virus

- **Virus:** single-stranded RNA virus.
    - Transmitted by exposure to contaminated blood; sexual transmission and mother-child transmission occur but are uncommon.
    - Hepatitis C virus (HCV) infects hepatocytes; hepatocyte destruction is mediated largely by the host immune response.
    - HCV RNA-dependent RNA polymerase lacks proofreading capacity leading to large variability in viral genome sequences; six biologically relevant genotypes are recognized although others likely exist (see *Genotyping Assays*).
    - HCV genotypes 1–3 are most common in North America and Europe.
        - US: Genotype 1 (Gt1), 75%; genotype 2 (Gt2), ~15%; genotype 3 (Gt3), ~5%.
- **Susceptible hosts/diseases:** Acute infection and the first decade(s) of chronic infection are largely asymptomatic. Progression from acute to chronic infection occurs in most individuals (~85%).
    - Spontaneous clearance and recovery from acute infection are associated with:
        - Female sex and young age.
        - Certain single nucleotide polymorphisms in the gene encoding interferon lambda (IL-28B, see *Genotyping Assays*).

- Disease progression during chronic infection:
  - Cirrhosis develops in 10%–20%.
  - Hepatocellular carcinoma occurs in ~5%.
- Recognized as an opportunistic infection in HIV. Incidence and rate of disease progression are greater in HIV/HCV coinfected individuals than in HCV monoinfected individuals.

Treatment

- The chronic hepatitis C treatment paradigm is response-guided therapy (RGT), or treatment for shortest period required to attain maximum cure rates.
  - RGT is based on observations that viremia clearance kinetics correlate with chronic hepatitis C cure rates (or sustained virologic response [SVR], defined as no detectable HCV RNA by a nucleic acid amplification test with a low limit of detection 6 months after treatment).
  - Patients who clear virus rapidly derive no additional benefit from extended therapy.
  - Improved SVR rates are obtained with longer treatment when HCV RNA clearance is prolonged.
  - On-treatment criteria ("futility rules") to identify non-responders have been established, permitting early drug discontinuation when no long-term benefit will be attained.
- Pegylated interferon alfa + ribavirin: Was the only available therapy for many years. Now used in the United States primarily for Gts 2, 3, and 4.
  - Treatment duration: Depending on response kinetics, 16–24 weeks treatment for Gt 2, 3; 24–72 weeks for Gt 4. Refer to the American Association for the Study of Liver Diseases [AASLD] 2009 Practice Guideline for further details and for use in genotype 1 infections.

  - Response rates: Gt 2, 75%–90%; Gt 3, 65%–75%; Gt 4, (40%–60%, comparable to Gt 1).
- HCV protease inhibitors (PIs) boceprevir and telaprevir: Developed for Gt 1 treatment; have minimal activity against Gt 2 and no activity against Gt 3 and 4. Can be used in treatment-naïve, previous treatment failure, compensated cirrhosis patients.
  - Used in combination with pegylated interferon alfa + ribavirin.
    - Telaprevir: All three drugs initiated at once.
    - Boceprevir: Lead-in phase with pegylated interferon alfa + ribavirin then addition of boceprevir.
  - SVR rates ~35% greater with triple drug PI-containing regimens than pegylated interferon alfa + ribavirin alone.
  - RGT rules vary by drug and prior treatment experience. Futility rules vary by drug relative to timing and virus concentration. Refer to the AASLD 2011 Practice Guideline for further details.
  - Improved response rates spurred initial treatment wave after which severe side effect profile of these agents became clear. The decision to start PI-based treatment is complex and multifactorial; the most critical factor is histologic severity of disease.
- Certain single nucleotide polymorphisms in the gene encoding interferon lambda (IL-28B) correlate with interferon-responsiveness (see *Genotyping Assays*).

Test Utility

- Distinguish resolved versus chronic infection in HCV seropositive patients; HCV RNA will be detectable in chronically infected individuals.
- Diagnosis of acute hepatitis C (in conjunction with serology).
- Monitoring therapeutic efficacy.
  - Testing performed prior to treatment initiation, on treatment (time points, RGT criteria, and futility rules are regimen-dependent. Refer to the 2009 and 2011 AASLD Practice

Guidelines for further details, end of treatment, and end of follow-up (6 months after treatment to assess SVR).

Tests

- HCV RNA qualitative detection assay based on transcription-mediated amplification (Aptima HCV, Hologic Gen-Probe) is approved for use in the diagnosis of chronic hepatitis C in individuals who have antibodies against HCV.
- HCV RNA quantification tests with low detection limits (10–15 IU/mL) and broad measurable ranges (10-15 IU/mL → 8.0 $\log_{10}$ IU/mL) should be used in therapeutic monitoring. Assays should be calibrated according to international standard and results reported as international units/mL. FDA-approved real-time PCR assays for use on automated platforms are available (Realtime HCV, Abbott Molecular; Cobas AmpliPrep/Cobas TaqMan HCV v2.0, Roche Molecular Diagnostics). These tests are labeled for use in therapeutic monitoring, but in practice are also diagnostically used.

Sample Types

Plasma ethylenediaminetetraacetic acid (EDTA), serum.

Limitations

- Sensitive nucleic acid detection (or quantification) tests should be used in conjunction with serologic tests for the diagnosis of acute hepatitis C.
  - Viremia can be intermittent and variably detected by nucleic acid tests after exposure and antibody production can be delayed.
  - HCV RNA detection in acute infection does not infer progression to chronic infection.
- Viral loads at end of treatment can be below the detection limits of current assays.

- Relapsed infections are observed in a minor proportion of patients with no detectable HCV RNA at end of treatment.
  - Individuals with no detectable HCV RNA at end of treatment should be tested six months later to ascertain SVR.

Human Immunodeficiency Virus

- **Virus:** Enveloped single-stranded RNA genome.
  - Transmitted via multiple routes including blood-borne and contact with secretions.
  - Two species, HIV-1 and HIV-2, cause disease globally.
    - The majority of infections are due to HIV-1; HIV-2 is generally limited to West Africa.
  - HIV infects CD4 cells and macrophages.
  - RNA genome is reverse-transcribed into double-stranded DNA form by viral reverse transcriptase.
    - Integration of the double-stranded viral genome into the host cell genome is mediated by viral integrase, forms the provirus, and leads to chronic infection.
    - Reverse transcriptase has no proofreading activity, is therefore error-prone, and leads to wide genetic variability, including mutations that confer antiviral resistance (see *Genotyping Assays*).
      - Recombination between strains is another source of genetic diversity.
    - HIV-1 is genotypically classified into three groups based on *gag* and *env* sequences: M (Major: Viruses that cause most infections worldwide), O (outlier), and N (non-M, non-O).
      - M: Worldwide distribution. Further subdivided into nine clades (A-D, F-H, J, K). Clade B is predominant in global epidemic; most infections in U.S. and Europe are clade C. Clade “E” is actually circulating recombinant form (CRF) between clades A and E prevalent in Southeast Asia.

        - O: Uncommon viruses; geographically restricted compared to M (Cameroon, Gabon, Equatorial Guinea)
        - N: Rare viruses, even greater geographic restriction than O viruses (Cameroon).

- **Susceptible hosts/diseases:**
    - Acute HIV-1 infection is generally symptomatic but can be mistaken for influenza due to similar symptoms or easily overlooked due to the nonspecific nature of symptoms. Patients typically present with:
        - Constitutional symptoms: Fever, myalgias, arthralgias.
        - Respiratory symptoms: Dry cough.
        - Gastrointestinal symptoms: Nausea, vomiting, diarrhea, anorexia.
        - Rash.
        - Painful ulcers (mouth, esophagus, rectum, anus, penis).
        - Laboratory findings: Leucopenia, thrombocytopenia, elevated serum alanine aminotransferase and aspartate aminotransferase levels.
    - In the earliest phase of acute infection, viremia is the only evidence of infection.
        - HIV-1 RNA detectable 6 days before p24 antigen and 12 days before antibody.
        - Screening for viremia by nucleic acid testing is effective in identifying acutely infected individuals.
    - Long periods of asymptomatic infection ensue in untreated individuals.
        - Eventually defective cell-mediated immunity occurs due to declining numbers of CD4 cells, predisposing individuals to opportunistic infections and cancers, resulting in high morbidity and mortality rates.
        - HIV viral load is one of several markers, including CD4 count to assess disease stage (Table 16).

- HIV-2 is less pathogenic than HIV-1; associated with longer asymptomatic period and slower progression to AIDS.

HIV-1 Antiretroviral Therapy (ART) Treatment

- Treatment is now recommended for any infected individual to reduce disease progression rates and prevent transmission. Refer to the US Department of Health and Human Services (DHHS) 2013 and International Antiviral Society (IAS) 2012 guidelines for more detail.
- Three regimens for use in treatment-naïve individuals are currently recommended (see the USDHHS 2013 and IAS 2012 guidelines):
  - Non-nucleoside reverse transcription inhibitor-based; or
  - Protease inhibitor-based; or
  - Integrase strand transfer inhibitor-based.

Test Utility

See Table 16.

Tests (US FDA-approved and other regulatory status assays)

- Qualitative detection, HIV-1 RNA: Currently approved test uses transcription-mediated amplification (Aptima HIV-1 Qualitative RNA Test, Gen-Probe). Limit of detection is ~30 copies/mL. Dual target amplification (long terminal repeat [LTR] and *pol*) allows detection of group M, N, and O viruses.
- Qualitative detection, HIV-1 RNA/DNA (Cobas AmpliPrep/Cobas TaqMan HIV-1 Qual, Roche Molecular Systems): Regulatory status is "Research Use Only."
  - Real-time PCR assay amplifies *gag* gene target; optimized for M group virus detection.
  - Pre-extraction cell lysis allows for amplification of proviral DNA and viral RNA in whole blood (EDTA) or dried blood spots samples.

Table 16. Uses for US FDA-Approved HIV-1 RNA Assays

Indication	*Approved Tests*	*Comment*
Diagnosis of acute infection in individuals with positive screening tests and negative or indeterminate confirmatory tests (acute infection)	RNA qualitative	• HIV-1 RNA detectable 12 days before antibody and 6 days before p24 antigen. • Guidelines note that quantitative assays can be used, but they are currently only US FDA-approved for baseline and on-treatment viral load measurement.
Confirmation of infection in individuals who have repeatedly reactive serology test results	RNA qualitative	• Could be used in infants born to mothers with unknown infection status at birth after HIV-1 is confirmed serologically in mother or baby. • In infants with known perinatal exposure.
Diagnosis of infection in infants with known perinatal exposure	RNA qualitative	• Guidelines recommend testing at 14–21 days, 1–2 months, and 4–6 months; positive results should be confirmed by testing second specimen. • Serology is inaccurate due to passively acquired maternal antibodies.

(*Continued*)

Table 16. Uses for US FDA-Approved HIV-1 RNA Assays (*Continued*)

Indication	*Approved Tests*	*Comment*
Baseline viral load at entry into HIV care to assess disease stage in patients with repeatedly positive serologic screening test and positive confirmatory testing	RNA quantification	• Early in disease: best predictor of outcome (rate of CD4 decline, time to AIDS, death over 10-y span); CD4 provides additional diagnostic accuracy. • Late (CD4 < 200/mm^3): predictive of opportunistic infection risk, independent of CD4 counts.
Monitoring ART efficacy (viral load is a marker of treatment response); see Table 17 for description of response definitions	RNA quantification	• Decreased viremia correlates with improved outcomes. • Testing time points: ○ First measurement should be 2–4 weeks after treatment initiation or change in regimen. Viral load should decrease by at least 10-fold. ○ Subsequent measurements every 4–8 weeks • Treatment goal is undetectable viremia after 16–24 weeks; then continue monitoring every 3–6 months.

- ○ Pediatric HIV guidelines have historically recommended proviral DNA amplification tests for diagnosis of infection in newborns (infants born to mothers with unknown infection status at birth after HIV-1 is confirmed serologically in mother or baby).

Table 17. HIV-1 Virologic Response Definitions as Applied to Therapeutic Monitoring

Virologic Response Terms	*Definition*
Virologic suppression	Confirmed[a] HIV-1 RNA < assay limit of detection
Virologic failure	Inability to reach or maintain viral load < 200 copies/mL
Incomplete virologic response	Confirmed HIV-1 RNA > 200 copies/mL after 24 weeks of treatment
Virologic rebound	Confirmed viral HIV-1 RNA > 200 copies/mL after achieving virologic suppression
Persistent low-level viremia	Confirmed viral HIV-1 RNA detectable, but < 1000 copies/mL
Virologic blip	Isolated detectable HIV-1 RNA after suppression

[a]Confirmed refers to re-testing (same specimen or new specimen, from any time) with the same assay.

    - Currently, direct HIV-1 RNA and DNA amplification tests are recommended; HIV-1 qualitative RNA test is the only test US FDA-approved for this use.
- Quantification, HIV-1 RNA: A number of US FDA-approved tests using a variety of chemistries are available (Table 18).
  - HIV-1 genetic variability has impacted nucleic acid quantification tests.
    - Group O and some group M viruses were not accurately quantified by first generation assays.
  - Currently approved qualitative and quantitative tests have improved genotype inclusivity.

Table 18. FDA-Approved HIV-1 RNA Quantification Tests

Test (Manufacturer)	*Chemistry*	*Measuring Range (Copies/mL)*	*Approved Groups (Subtypes)*
RealTime TaqMan HIV-1 (Abbott Molecular)	Reverse transcription/real-time PCR (integrase sequences within *pol* gene)	40–10,000,000	M (A-H), O, N
NucliSens HIV-1 QT (BioMerieux)	NASBA, *gag* gene	176–3,470,00	M (A-H)
Cobas AmpliPrep/Cobas TaqMan HIV-1 v1 (Roche Diagnostics)	Reverse transcription/real-time PCR (*gag* gene)	48–10,000,000	M (A-H)
Cobas AmpliPrep/Cobas TaqMan HIV-1 v2 (Roche Diagnostics)	Reverse transcription/real-time PCR (*gag* gene and LTR)	20–10,000,000	M (A-H, CRFs), O, N
Versant HIV-1 RNA (Siemens Healthcare Diagnostics)	bDNA (*pol* gene)	75–500,000	M (A-G)

LTR, long terminal repeat; NASBA, Nucleic acid sequence-based amplification; PCR, polymerase chain reaction.

Sample Types

- Qualitative detection, HIV-1 RNA: Plasma or serum.
- Qualitative detection, HIV-1 RNA/DNA: Plasma, whole blood (EDTA), dried blood spots.
- Quantification, HIV-1 RNA: Plasma (EDTA or acid citrate dextrose [ACD], consult product inserts) or serum.

Limitations

- HIV-1 RNA quantification assays do not amplify HIV-2. Separate assays must be used for HIV-2 RNA measurement; no US FDA-approved HIV-2 RNA quantification tests are available.
- For HIV-1 quantification assays, viral load changes $\geq 0.5\log_{10}$ copies/mL are considered biologically significant.
  - Viral load can vary up to 0.3 $\log_{10}$ copies/mL over time in untreated patients.
  - Assay precision can vary up to 0.3 $\log_{10}$ copies/mL depending upon the assay and the viral load (in real-time PCR assays, imprecision at low viral loads > high viral loads).
- Clinical significance of virologic blips and viral loads < 200 copies/mL is controversial.

Parvovirus B19

- **Virus:** Nonenveloped, single-stranded DNA genome; infects erythroid progenitor cells.
- **Susceptible hosts/diseases:**
  - Aplastic crisis (chronic hemolytic anemia patients with a mandatory physiologic requirement for constant red cell production due to continuous red cell destruction).
  - Chronic anemia (immunocompromised patients, especially HIV; chronic anemia due to inability to clear viral infection).
  - Hydrops fetalis (fetal infection).

- Fifth disease (children, usually diagnosed clinically or serologically).
- Arthropathy (adults; women>men).
- Myocarditis (immunocompetent patients of any age, both genders).
- Palpable-purpuric stocking-glove rash (rare, adults; women>men).

Test Utility

Parvovirus B19 does not cause asymptomatic or latent infection; therefore, nucleic acid detection is sufficient for diagnosis of active, disease-producing infection.

Tests

- Qualitative nucleic acid detection is useful; quantitative tests have also been adopted.
- No US FDA-approved tests are available. Most assays are laboratory-developed real-time PCR tests.

Sample Types

- Serum/plasma (for anemia, arthropathy, myocarditis, acute maternal infection).
- Amniotic fluid (for cases of hydrops fetalis).

Limitations

- International standards for use as calibrators to standardize quantification are available (second World Health Organization International Standard for parvovirus B19 DNA for nucleic acid testing assays). Although this preparation is available, most quantitative assays are not standardized despite the availability of this preparation and results are commonly reported in copies/mL.

- Interassay variability in quantification is expected given the use of varying calibrators in different tests; patients should be monitored with a single assay.
- Quantitative cutoffs to guide management are not available.

Genotyping Assays for the Management of Infectious Diseases

Genotype Testing in Cytomegalovirus Infections

Detection of Antiviral Drug-Resistance Mutants

Ganciclovir is the initial drug of choice for prophylaxis (in D+/R– allograft recipients), preemptive treatment of viremia, and disease.

- Ganciclovir triphosphate is a suicide substrate for the viral DNA polymerase (encoded by virus gene *UL54*). The drug is mono- and diphosphorylated by cellular kinases; triphosphorylation is performed by a viral kinase encoded by virus gene *UL97*.
- Drug resistance emerges during treatment as result of drug selection. Wild type viruses are not inherently resistant and resistant viruses are not thought to be transmitted.
- *UL97* mutations most commonly cause ganciclovir resistance. *UL54* mutations usually arise after *UL97* mutations, produce high-level resistance (high drug concentrations required to inhibit 50% of virus growth *in vitro*, "IC50"), and confer cross-resistance to other polymerase-active drugs (foscarnet and/or cidofovir).
- Resistance occurs most commonly in solid organ transplant recipients. Risk factors include prolonged drug treatment (usually ganciclovir) and causes of high-level replication including lack of immunity (CMV seronegative recipients of allografts from seropositive donors, "D+/R–") and intensive immunosuppression.
- Testing is indicated in individuals with prolonged antiviral treatment history and evidence of viremia rebound, continued high-level viremia, or progressive disease.

Tests

- Direct sequencing is clinically most practical; phenotypic assays (virus growth reduction in the presence of drug) are used to determine significance of uncharacterized mutations and assign resistance level (high or low IC_{50}).
  - *UL97* kinase gene (codons 400–670) is usually first sequencing target.
  - *UL54* gene sequencing (codons 300–1000) should be considered if resistance to high-dose ganciclovir or resistance to polymerase-active agents (foscarnet, cidofovir) suspected.
- Direct sequencing results are reported as mutation(s) known to cause resistance, single nucleotide polymorphisms with no associated drug-resistance phenotype, or previously unreported sequence change with unknown phenotype after referring to the most current, comprehensive list of sequence changes.

Sample Types

Plasma, serum.

Limitations

- Plasma viral loads of ~500–1000 IU/mL are required for direct sequencing.
- False-negative results may be obtained if drug-resistant variants are $\leq$ 10%–20% of circulating virus due to the inability of direct sequencing to detect low prevalence minority variants.

Genotype Testing for the Management of Chronic Hepatitis C Virus Infection

HCV Genotype/Subtype

- Six genotypes (Gt 1–6) with variable global distribution. Dominant genotypes:

- US and Europe—Gt 1-3 (Gt 1, 75%; Gt 2, 15%; Gt 3, 5%)
- Egypt—Gt 4
- South Africa—Gt 5
- Asia—Gt 6

Each genotype has multiple subtypes. The only clinically relevant subtypes currently are Gt 1a and 1b.

- Genotype utility: Genotype is a critical viral marker in management; currently the primary determinant of treatment regimen selection. Its importance may change since pan-genotypic, directly acting antiviral drugs are in development. See *Infections Managed Through Nucleic Acid Testing of Blood/HCV* for genotype-specific treatment regimens and response rates.
- Gt 1 subtype utility (Gt 1a vs Gt 1b): Resistance rates to the PIs boceprevir and telaprevir are subtype-dependent (Gt 1a > Gt 1b). Resistance conferred by single nucleotide change in Gt 1a; two nucleotide changes required in Gt 1b.
    - Subtype is a factor in the complex, multifactorial decision to initiate or defer treatment with boceprevir and telaprevir-based regimens.
    - In some Gt 1a patients, treatment deferment is advised to minimize development of PI-resistant strains that might not be responsive to future higher potency, better-tolerated PI-based regimens.

Tests

- No US FDA-approved tests available.
- Two commercial tests ("Research Use Only") include reverse hybridization of 5′ UTR and core amplicons to oligonucleotides immobilized on membranes (line probe) and direct (population-based) 5′ UTR sequencing.
- Direct (population-based) sequencing laboratory developed tests (core-E1 and core/NS5b) are also in use.

Sample Types

Plasma, serum.

Limitations

Assays that characterize only 5'UTR sequence can yield inaccurate Gt 1 subtype results due to sequence conservation; improved subtyping accuracy is obtained with core (or core/E1) or NS5b sequence analysis.

HCV Genotype for Resistance to Direct-Acting Antiviral Drugs

Two direct-acting antiviral (DAA) drugs, the NS3/4a PIs boceprevir and telaprevir have been approved for treatment of individuals chronically infected with HCV Gt 1. The HCV RNA genome is highly mutable and antiviral resistance occurs rapidly after protease monotherapy. Therefore, current regimens consist of DAAs plus pegylated interferon alfa and ribavirin. Mutations in the NS3/4a catalytic site confer resistance by impairing drug binding. Protease resistant viruses are found in ~90% of patients with virologic failure to triple therapy. Despite this, there are currently no recommendations for resistance detection through genotyping, based on the following:

- Mutations are detected in ~5% of individuals at baseline and their presence is not an indicator of treatment outcome. Detection of PI resistant variants prior to treatment initiation is therefore not warranted.
- During treatment, inability to respond to interferon alfa is associated with a higher risk of PI resistance, suggesting that inadequate immunologic suppression of viral replication allows for the selection or emergence of resistant mutants. Currently, prevention of resistance during treatment entails viral load monitoring for early identification and treatment cessation in interferon non-responders. The use of genotype tests to manage the emergence of PI resistant viruses during treatment has not been recommended.

- The detection of resistant viruses should not be used to switch from one PI to the other during treatment since cross-resistance is observed.

Tests

Methods to detect resistant variants include direct, population-based sequencing, next generation sequencing, and sequencing individual viral clones.

Sample Types

Plasma, serum.

Limitations

PI resistance detection has no current clinical utility; testing is largely performed for research purposes.

Determination of Host Genotype for the Management of Chronic Hepatitis C Virus Infection

- Two single nucleotide polymorphisms (SNPs) on chromosome 19 near the gene encoding IL28B (lambda interferon 3) are associated with clearance of acute infection and attainment of sustained virologic response.
  - Associations are stronger for Gt1 infections compared with Gt 2 and Gt 3. The SNPs (alleles) are rs12979860 (C or T) and rs809917 (T or G).
- Genotypes associated with highest rates of spontaneous virus clearance after acute infection and sustained virologic response to pegylated interferon/ribavirin and boceprevir or telaprevir plus pegylated interferon/ribavirin (triple therapy):
  - rs12979860 CC
  - rs809917 TT

- Spontaneous virus clearance and sustained virologic response rates are lower for:
  - rs12979860 CT and TT genotypes
  - rs809917 TG and GG genotypes
- Current guidelines recommend considering host genotype determination in previously untreated individuals chronically infected with HCV Gt 1 to provide information regarding likelihood of response and treatment duration.
  - Genotype is not recommended to guide selection of pegylated interferon alfa/ribavirin over triple therapy in CC individuals. Triple therapy is preferable since response rates appear to be higher and treatment duration may be shortened compared with pegylated interferon alfa/ribavirin alone.
  - Genotype is not recommended as rationale for withholding treatment from individuals with unfavorable genotypes; approximately one-half of these attain sustained virologic response to triple therapy.
  - Genotype is not informative for Gt 1 individuals who failed therapy with pegylated interferon alfa/ribavirin since virtually all have unfavorable genotypes.

Tests

- No US FDA-approved tests are available.
- Laboratory-developed methods include direct sequencing (Sanger, pyrosequencing) and real-time PCR with various probe-strategies for SNP discrimination.

Sample Types

Whole blood.

Limitations

IL28B and adjacent genes *IL28A* and *IL29* are highly homologous. Sequence characterization assays must be well designed to unambiguously determine IL28B SNPs.

Genotype Testing for the Management of Chronic Hepatitis B Virus Infection

HBV Genotype

Eight genotypes (A–H) with distinct geographic distributions have been identified.

- Gt A most prevalent in the United States
- Response rates to pegylated interferon treatment:
  - Gt A > Gt D among Caucasians
  - Gt B > Gt C among Asians

Tests

- No US FDA-approved tests available.
- Direct sequencing (commercial "Research Use Only" assay and laboratory-developed tests) and a commercial reverse hybridization test (viral PCR products hybridized to membrane-bound oligonucleotides, "Research Use Only") can be used.
- Surface antigen coding sequence is characterized in all test formats.

Sample Types

Plasma, serum.

Limitations

There are currently no recommendations to use genotype as the basis for interferon treatment since response likelihood varies among individuals.

Basal Core Promoter/Precore Mutations

These mutations abrogate HBeAg expression, but can still result in chronic hepatitis B (cHB).

- Individuals with HBeAg-negative cHB can have normal serum transaminase levels and low HBV viremia making them difficult to distinguish from inactive carriers who share the same serologic profile (HBsAg positive/negative, HBeAg negative/positive, hepatitis B core IgG positive, hepatitis B core IgM negative, HBV DNA detectable). See *Infections Managed Through Nucleic Acid Testing of Blood /HBV* for additional information.
- HBeAg negative cHB usually diagnosed by serial viral load assessments; HBV DNA > 2000 IU/mL is a good indicator. It can also be diagnosed using assays that detect basal core promoter/precore mutations.

Tests

- No US FDA-approved tests available.
- Direct sequencing (laboratory developed tests) and a commercial reverse hybridization assay ("Research Use Only") are used.

Sample Types

Plasma, serum.

Limitations

- Although genotypic characterization of these mutations is feasible, testing is not readily accessible.
- The diagnosis of HBeAg negative cHB is usually made through serial quantitative HBV DNA assessment.

Hepatitis B Virus Genotyping for Antiviral Drug-Resistance Mutants

- Drug resistance emerges during treatment as result of drug selection.
- Resistance rates are drug dependent (lamivudine 70% at 5 y, adefovir, 29% at 5 y, emtricitabine 18% at 2 y, telbivudine 17% at 2 y, entecavir 1.2% at 5 y, tenofovir 0% at 4.5 y).
- Drug-resistance testing is indicated when viremia rebounds during treatment to prove resistance versus lack of compliance and to guide selection of alternate therapeutic options.

Tests

- Two common assay formats for polymerase gene sequence analysis:
    - Population-based direct sequencing (commercial assay and lab developed tests).
    - Reverse hybridization of PCR products to oligonucleotides immobilized on membranes.
        - Hybridization-based test is sensitive, can detect mixed populations, and low prevalence (> 5%) variants.

Sample Types

Plasma, serum.

Limitations

- Direct sequencing: New drug-resistance mutations are readily detectable; however, their significance can be difficult to interpret.
    - Impact on drug resistance requires clinical correlation (association with rebound viremia in a number of patients) or phenotypic confirmation.
    - False-negative results may occur if mutations are minority variants (constituting < 10%–20% of virus population).

- Reverse hybridization: Interpretation can be complex due to the large number of bands (34) and to occurrence of faint bands. Use of an automated reader can be helpful.
  - False-negative results can arise from lack of PCR amplification.
  - New resistance mutations not detected with this method.

Vaccine-Escape Hepatitis B Virus Surface Antigen Mutations

- A single amino acid change in HBsAg (glycine to arginine at codon 145) inhibits neutralization of HBV infection by vaccine-induced hepatitis B surface antibody.
  - Infections with these mutant viruses can produce false-negative HBsAg results by some serologic assays, leading to misdiagnosis of HBV infection.
  - Transmission to vaccinated individuals can occur as immunization does not protect against infection.
  - The prevalence of these viruses initially increased after initiation of vaccination programs in Asia suggesting that they arise due to immune selection. However, prevalence rates have stabilized recently and HBV vaccine efficacy rates remain high globally.

Tests

Mutants can be definitively detected by direct, population-based S gene sequencing.

Sample Types

Plasma, serum.

Limitations

- These viruses are uncommon in the United States and genotypic characterization is not readily accessible.

- The diagnosis can be made through the use of HBsAg assays that can detect these mutant viruses and other infection markers (hepatitis B core IgG positive and HBV DNA detectable).

Genotype Testing in the Management of Human Immunodeficiency Virus Type 1 Infections

Detection of Antiviral Drug-Resistance Mutants

The major drug classes for the treatment of HIV-1 are reverse transcriptase (RT) inhibitors, protease (PR) inhibitors, integrase (INT) inhibitors, and viral entry inhibitors.

- Resistance to any of these agents can be conferred by mutations in drug target genes.
- Resistant strains can be transmitted from person to person and can emerge during drug treatment.
  - Rates of resistance to PR inhibitors are low; rates for RT and INT inhibitors are higher.

Assays for the detection and characterization of drug-resistance mutations in HIV-1 infections are recommended in the following settings:

- At diagnosis to characterize the acquired virus and assist in drug regimen selection.
  - Rates of drug-resistant HIV-1 transmission vary with regional access to ART, transmission risk factor, and duration of infection prior to testing.
- Before treatment initiation to detect any drug-resistant viruses acquired after diagnosis.
- During treatment, to investigate virologic failure.
- During pregnancy, to assist with selection of appropriate therapy that is safe for the fetus but will suppress replication in mother and prevent vertical transmission.

Tests

Three assay formats are available.

- Genotyping assays: Population-based direct sequencing for RT and PR mutations only.
  - Laboratories must have up-to-date libraries since new resistance mutations are continuously identified.
  - These tests more effectively identify virus mixtures than phenotyping assays.
- Phenotype assays: Viral genes from clinical samples are inserted into HIV-1 vectors. Replication is measured as function of reporter gene expression.
  - Test is performed in presence of varying drug concentrations.
  - Drug concentration that results in 50% inhibition of viral replication (IC_{50}) is calculated and compared with wild type.
  - Results are reported in reference to biologic cutoffs designating partial and complete resistance based on treatment outcome.
- Virtual phenotyping assay: Genotyping by direct, population-based sequencing; a database containing mutations paired with phenotypes is then queried to interpret results.
  - Individual laboratories can perform direct sequencing then access database or plasma can be sent to commercial lab for direct sequencing plus virtual phenotype database analysis.
  - Helpful when phenotyping is necessary but not available.

Limitations

- Genotyping assays:
  - Individuals with low viral loads (persistent low level viremia) may be difficult to characterize since plasma viral loads of 500–1000 copies/mL are required for direct sequencing.
  - Interpretation of genotype data can be difficult in individuals with extensive treatment experience due to large number of mutations associated with resistance to multiple drugs.
    - Genotyping is therefore useful for treatment-naïve individuals or those with limited treatment history.

    - Genotype plus phenotype may be useful in treatment-experienced individuals with evidence of virologic failure.
  - Low prevalence minority variants that ultimately cause virologic failure may not be effectively detected. False-negative results may be obtained if variants are < 10%–20% of circulating virus.
  - Genotyping can fail to detect drug-resistant transmitted strains if treatment initiation is substantially deferred.
  - Drug-resistant virus mutants can have diminished fitness compared to drug susceptible strains.
    - In untreated patients (in the absence of drug selection), these mutants may disappear from circulation.
    - After treatment initiation, mutant viruses can selectively emerge from lymphoid reservoirs causing virologic failure.
- Phenotyping assays:
  - Time to result can be slower than genotyping.
  - Available as send-out to commercial laboratories only.

Characterization of HIV-1 Tropism

- HIV-1 virions enter host cells through the following sequence of events:
  - Viral gp120 protein engages the cellular receptor CD4 and undergoes a conformational change that allows gp120 to bind the host cell co-receptor, one of two chemokines, CCR5 or CXCR4. CCR5 is found on macrophages and CXCR4 is found on T cells.
  - Chemokine co-receptor engagement induces a conformational change in the viral protein gp41 and exposes a hydrophobic domain on the molecule that leads to fusion of the virion envelope with the host cell plasma membrane.
- HIV-1 strains that utilize CCR5 are designated macrophage-tropic and predominate early in infection; strains that utilize CXCR4 are T cell-tropic and emerge later. Some HIV-1 strains are dual-tropic (can use either coreceptor for entry).

- The HIV-1 entry inhibitor maraviroc blocks the chemokine coreceptor CCR5 and inhibits infection by macrophage-tropic strains. It does not block CXCR4 and is not effective for the treatment of infections with T cell-tropic or dual tropic HIV-1.
- The co-receptor usage of an individual's virus must be determined prior to maraviroc initiation to determine whether treatment will be effective.
    - Maraviroc is currently recommended as an alternate drug for use after virologic failure with first-line regimens.

Tests

Two test formats are available for tropism determination:

- *env* sequences amplified from patient plasma expressed on surface of luciferase-expressing hybrid virus ("pseudotype") that is then used to infect CD4+ cells expressing either CXCR4 or CCR5.
- Heteroduplex mobility assay plus V3 loop sequencing. Heteroduplexes between patient *env* sequences and *env* from known CXCR4 and CCR5 detected by capillary electrophoresis.

Sample Types

Plasma.

Limitations

Tropism determination is highly specialized testing that is offered only by reference laboratories.

High-Risk Human Papillomavirus Genotype Determination

- The majority of cervical cancers worldwide are caused by HPV types 16, 18, and 45 (approximately 60%, 10% and 5%; Table 19).
- In women ≥ 30 y of age who were cotested (Pap smear cervical cytology plus high-risk HPV qualitative detection) and genotyped if

found to be HR HPV positive, the 10-y cumulative incidence rates of invasive cancer were approximately 20% if HPV 16 or HPV 18 were present. Given these data, current guidelines recommend the following in cotested women with negative Pap smear and positive HR HPV results:

- Repeat cotesting in 12 months; or
- Immediate genotyping for HPV 16/18.
  - If HPV 16 or 18 positive → direct referral to colposcopy
  - If HPV 16 or 18 negative → repeat co-testing at 12 months

- Cotested women ≥ 30 y of age with negative Pap smear and positive HR HPV results SHOULD NOT be referred directly to colposcopy unless indicated by genotype test results, due to the overall low risk of invasive cervical cancer within a 1-y period of time.

Tests

A number of FDA-approved tests using a variety of chemistries are available (Table 20).

Table 19. Prevalence of HPV 16/18/45 in Cervical Cancers

Tumor Type	*Tumors Caused by Different HPV Types (%)*		
	HPV 16	**HPV 18**	**HPV 45**
Squamous cell carcinoma	62	8	5
Adenocarcinoma	50	32	12
Adenosquamous cell carcinoma	39	32	12

From deSanjose S, Quint WGV, Alemany L, et al. Human papillomavirus genotype attribution in invasive cervical cancer: a retrospective cross-sectional worldwide study. Lancet Oncol 2010;11:1048–56.

Table 20. FDA-Approved Tests for High-Risk HPV Genotyping

Test (Manufacturer)	*Detection Chemistry*	*HPV Types Detected*	*Comment*
Aptima HPV (Hologic Gen-Probe)	Transcription mediated amplification	16, 18, 45	Separate tests required for HR HPV detection and genotyping. Inclusion of HPV 45 allows identification of additional HR type associated with cervical cancer however current guidelines have no recommendations on management of women >30 y with negative Pap smear, positive HR HPV and HPV 45.
Cervista HPV HR (Hologic)	Invader (target amplification)	16, 18	Separate tests required for HR HPV detection and genotyping.
Cobas 4800 HPV (Roche Molecular Diagnostics)	Real-time PCR	16, 18	Multiplex format; detects HR HPVs and individually identifies types 16 and 18 in a single test.

HR HPV, high-risk human papillomavirus; PCR, polymerase chain reaction.

Sample Types

Liquid-based cervical cytology medium (PreservCyt) has been approved for use in these assays.

Limitations

- Consensus committees have found no compelling evidence supporting the clinical utility of HR HPV genotyping other than in cotested women > 30 y of age with negative Pap smear and positive HR HPV results.
    - Current guidelines therefore do not recommend any other uses.
- The cumulative 1-y risk of invasive cervical cancer for types other than HPV 16 and 18 are low enough to substantiate recommendations for 12-month follow-up with repeat cotesting.

Suggested Reading

Allen U, Preiksaitis J; AST Infectious Diseases Community of Practice. Epstein-barr virus and posttransplant lymphoproliferative disorder in solid organ transplant recipients. Am J Transplant 2009;(9 Suppl 4):S87–96.

ALTS Group. Results of a randomized trial on the management of cytology interpretations of atypical squamous cells of undetermined significance. Am J Obstet Gynecol 2003;188:1383–92.

Andrea SB, Chapin KC. Comparison of Aptima Trichomonas vaginalis transcription-mediated amplification assay and BD Affirm VPIII for detection of T. vaginalis in symptomatic women: performance parameters and epidemiological implications. J Clin Microbiol 2011;49:866–9.

Boivin G. Diagnosis of herpesvirus infections of the central nervous system. Herpes 2004;(11 Suppl 2):48A–56A.

Brew BJ, Davies NW, Cinque P, et al. Progressive multifocal leukoencephalopathy and other forms of JC virus disease. Nat Rev Neurol 2010;6:667–9.

Calfee DP, Salgado CD, Classen D, et al. Strategies to prevent transmission of methicillin-resistant Staphylococcus aureus in acute care hospitals. Infect Control Hosp Epidemiol 2008;(29 Suppl 1):S62–80.

Carroll KC, Bartlett JG. Biology of Clostridium difficile: implications for epidemiology and diagnosis. Annu Rev Microbiol 2011;65:501–21.

Castle PE, Schiffman M, Burk RD, et al. Restricted criss-reactivity of hybrid capture 2 with non-oncogenic human papillomavirus types. Cancer Epidemiol Biomarkers Prev 2002;11:1394–9.

Chen CJ, Yang HI, Su J, et al; REVEAL-HBV Study Group. Risk of hepatocellular carcinoma across a biological gradient of serum hepatitis B virus DNA level. JAMA 2006;295:65–73.

Chotiyaputta W and Lok AS. Hepatitis B virus variants. Nat Rev Gastroenterol Hepatol 2009;6:453–62.

Cohen MS, Chen YQ, McCauley M, et al. Prevention of HIV-1 infection with early antiretroviral therapy. N Engl J Med 2011;365:493–505.

Cohen SH, Gerding DN, Johnson S, et al. Clinical practice guidelines for Clostridium difficile infection in adults: 2010 update by the society for healthcare epidemiology of America (SHEA) and the infectious diseases society of America (IDSA). Infect Control Hosp Epidemiol 2010;31:431–55.

Cooley LA, Lewin SR. HIV-1 cell entry and advances in viral entry inhibitor therapy. J Clin Virol 2003;26:121–32.

Cosgrove SE. The relationship between antimicrobial resistance and patient outcomes: mortality, length of hospital stay, and health care costs. Clin Infect Dis 2006;(42 Suppl 2):S82–9.

Crobach MJ, Dekkers OM, Wilcox MH, et al. European Society of Clinical Microbiology and Infectious Diseases (ESCMID): data review and recommendations for diagnosing Clostridium difficile-infection (CDI). Clin Microbiol Infect 2009;15:1053–66.

deSanjose S, Quint WGV, Alemany L, et al. Human papillomavirus genotype attribution in invasive cervical cancer: a retrospective cross-sectional worldwide study. Lancet Oncol 2010;11:1048–56.

Dijkmans AC, de Jong EP, Dijkmans BA, et al. Parvovirus B19 in pregnancy: prenatal diagnosis and management of fetal complications. Curr Opin Obstet Gynecol 2012;24:95–101.

Echavarria M. Adenoviruses in immunocompromised hosts. Clin Microbiol Rev 2008;21:704–15.

Eid AJ, Arthurs SK, Deziel PJ, et al. Emergence of drug-resistant cytomegalovirus in the era of valganciclovir prophylaxis:therapeutic implications and outcomes. Clin Transplant 2008;22:162–70.

Einstein MH, Martens MG, Garcia FA, et al. Clinical validation of the Cervista HPV HR and 16/18 genotyping tests for use in women with ASC-US cytology. Gynecol Oncol 2010;118:116–22.

European Association for the Study of the Liver. EASL clinical practice guidelines: management of chronic hepatitis B virus infection. J Hepatol 2012;57:167–85.

European Association for the Study of the Liver. EASL clinical practice guidelines: management of hepatitis C virus infection. J Hepatol 2011;55:245–64.

Fiebig EW, Wright DJ, Rawal BD, et al. Dynamics of HIV viremia and antibody seroconversion in plasma donors: implications for diagnosis and staging of primary HIV infection. AIDS 2003;17:1871–9.

Florea AV, Ionescu DN, Melhem MF. Parvovirus B19 infection in the immunocompromised host. Arch Pathol Lab Med 2007;131:799–804.

French GL. Methods for screening for methicillin-resistant Staphylococcus aureus carriage. Clin Microbiol Infect 2009;15(Suppl 7):10–6.

Gartner B, Preiksaitis JK. EBV viral load detection in clinical virology. J Clin Virol 2010;48:82–90.

Gotuzzo E. Xpert MTB/RIF for diagnosis of pulmonary tuberculosis. Lancet Infect Dis 2011;11:802–3.

Green M, Michaels MG. Epstein-Barr virus infection and posttransplant lymphoproliferative disorder. Am J Transplant 2013;(13 Suppl 3):41–54.

Gulley ML, Tang W. Using Epstein-Barr viral load assays to diagnose, monitor, and prevent posttransplant lymphoproliferative disorder. Clin Microbiol Rev 2010;23:350–66.

Ghany MG, Nelson DR, Strader DB, et al. An update on the treatment of genotype 1 chronic hepatitis C virus infection: 2011 practice guideline by the American Association for the Study of Liver Diseases. Hepatology 2011;54:1433–44.

Ghany MG, Strader DB, Thomas DL, et al. American Association for the Study of Liver Diseases. Diagnosis, management, and treatment of hepatitis C: an update. Hepatology 2009;49:1335–74.

Goegebuer T, Van Meensel B, Beuselinck K, et al. Clinical predictive value of real-time PCR quantification of human cytomegalovirus DNA in amniotic fluid samples. J Clin Microbiol 2009;47:660–5.

Halfon P, Sarrazin C. Future treatment of chronic hepatitis C with direct acting antivirals: is resistance important? Liver Int 2012;(32 Suppl 1):79–87.

Harbarth S, Hawkey PM, Tenover F, et al. Update on screening and clinical diagnosis of methicillin-resistant Staphylococcus aureus (MRSA). Int J Antimicrob Agents 2011;37:110–7.

Henrich TJ, Kuritzkes DR. HIV-1 entry inhibitors: recent developments and clinical use. Curr Opin Virol 2013;3:51–7.

Hewlett EL, Edwards KM. Clinical practice. Pertussis--not just for kids. N Engl J Med 2005;352:1215–22.

Hirsch HH, Randhawa P; AST Infectious Diseases Community of Practice. BK virus in solid organ transplant recipients. Am J Transplant 2009;(9 Suppl 4):S136–46.

Hirsch HH, Brennan DC, Drachenberg CB, et al. Polyomavirus-associated nephropathy in renal transplantation: interdisciplinary analyses and recommendations. Transplantation 2005;79:1277–86.

Hongthanakorn C, Chotiyaputta W, Oberhelman K, et al. Virological breakthrough and resistance in patients with chronic hepatitis B receiving nucleos(t)ide analogues in clinical practice. Hepatology 2011;53:1854–63.

Huletsky A, Giroux R, Rossbach V, et al. New real-time PCR assay for rapid detection of methicillin-resistant Staphylococcus aureus directly from specimens containing a mixture of staphylococci. J Clin Microbiol 2004;42:1875–84.

Ison MG, Green M. AST Infectious Diseases Community of Practice. Adenovirus in solid organ transplant recipients. Am J Transplant 9 2009;(Suppl 4):S161–5.

Katki HA, Kinney WK, Fetterman B, et al. Cervical cancer risk for women undergoing concurrent testing for human papillomavirus and cervical cytology: a population-based study in routine clinical practice. Lancet Oncol 2011;12:663–72.

Khan MJ, Castle PE, Lorincz AT, et al. The elevated 10-year risk of cervical precancer and cancer in women with human papillomavirus (HPV) type 16 or 18 and the possible utility of type-specific HPV testing in clinical practice. J Natl Cancer Inst 2005;97:1072–9.

Khattab MA, Ferenci P, Hadziyannis SJ, et al. Management of hepatitis C virus genotype 4: recommendations of an international expert panel. J Hepatol 2011;54:1250–62.

Kimura H, Ito Y, Suzuki R, et al. Measuring Epstein-Barr virus (EBV) load: the significance and application for each EBV-associated disease. Rev Med Virol 2008;18:305–19.

Kotton CN, Kumar D, Caliendo AM, et al; Transplantation Society International CMV Consensus Group. International consensus guidelines on the management of cytomegalovirus in solid organ transplantation. Transplantation 2010;89:779–95.

Kraft CS, Armstrong WS, Caliendo AM. Interpreting quantitative cytomegalovirus DNA testing: understanding the laboratory perspective. Clin Infect Dis 2012;54:1793–7.

Lemonovich TL, Watkins RR. Update on cytomegalovirus infections of the gastrointestinal system in solid organ transplant recipients. Curr Infect Dis Rep 2012;14:33–40.

Lautenschlager I, Razonable RR. Human herpesvirus-6 infections in kidney, liver, lung, and heart transplantation: review. Transpl Int 2012;25:493–502.

Liaw YF, Leung N, Kao JH, et al. Chronic Hepatitis B Guideline Working Party of the Asian-Pacific Association for the Study of the Liver. Asian-Pacific consensus statement on the management of chronic hepatitis B: a 2008 update. Hepatol Int 2008;2:263–83.

Lisboa LF, Asberg A, Kumar D, et al. The clinical utility of whole blood versus plasma cytomegalovirus viral load assays for monitoring therapeutic response. Transplantation 2011;91:231–6.

Lok AS, McMahon BJ. Chronic hepatitis B. Hepatology 2007; 45:507–39.

Lok AS, McMahon BJ. Chronic hepatitis B: update 2009. Hepatology 2009;50:661–2.

Lok AS, Zoulim F, Locarnini S; Hepatitis B Virus Drug Resistance Working Group. Antiviral drug-resistant HBV: standardization of nomenclature and assays and recommendations for management. Hepatology 2007;46:254–65.

Lurain NS, Chou S. Antiviral drug resistance of human cytomegalovirus. Clin Microbiol Rev 2010;23:689–712.

Mahony JB. Detection of respiratory viruses by molecular methods. Clin Microbiol Rev 2008;21:716–47.

Mahony JB. Nucleic acid amplification-based diagnosis of respiratory virus infections. Expert Rev Anti Infect Ther 2010;8:1273–92.

Martin-Iguacel R, Llibre JM, Nielsen H, et al. Lymphogranuloma venereum proctocolitis: a silent endemic disease in men who have sex with men in industrialised countries. Eur J Clin Microbiol Infect Dis 2010;29:917–25.

Matthes-Martin S, Feuchtinger T, Shaw PJ, et al. Fourth European Conference on Infections in Leukemia. European guidelines for diagnosis and treatment of adenovirus infection in leukemia and stem cell transplantation: summary of ECIL-4 (2011). Transpl Infect Dis 2012;14:555–63.

Mazuski JE. Vancomycin-resistant enterococcus: risk factors, surveillance, infections, and treatment. Surg Infect (Larchmt) 2008;9:567–71.

Miller GG, Dummer JS. Herpes simplex and varicella zoster viruses: forgotten but not gone. Am J Transplant 2007;7:741–7.

Muto CA, Jernigan JA, Ostrowsky BE, et al. SHEA guideline for preventing nosocomial transmission of multidrug-resistant strains of Staphylococcus aureus and enterococcus. Infect Control Hosp Epidemiol 2003;24:362–86.

Panel on Antiretroviral Guidelines for Adults and Adolescents. Guidelines for the use of antiretroviral agents in HIV-1 infected adults and adolescents. Department of Health and Human Services. http://aidsinfo.nih.gov/ConentFiles/AdultandAdolescentGL.pdf (Accessed March 14, 2013).

Panel on Antiretroviral Therapy and Medical Management of HIV-infected Children. Guidelines for the use of antiretroviral agents in pediatric HIV infection. http://aidsinfo.nih.gov/contentfiles/lvguidelines/pediatricguidelines.pdf (Accessed March 14, 2013).

Parker A, Bowles K, Bradley JA, et al; Haemato-oncology Task Force of the British Committee for Standards in Haematology and British Transplantation Society. Diagnosis of post-transplant lymphoproliferative disorder in solid organ transplant recipients - BCSH and BTS Guidelines. Br J Haematol 2010;149:675–92.

Pilcher CD, Fiscus SA, Nguyen TQ, et al. Detection of acute infections during HIV testing in North Carolina. N Engl J Med 2005; 352:1873–83.

Ramirez MM, Mastrobattista JM. Diagnosis and management of human parvovirus B19 infection. Clin Perinatol 2005;32:697–704.

Razonable RR. Human herpesviruses 6, 7 and 8 in solid organ transplant recipients. Am J Transplant 2013;(13 Suppl 3):67–77.

Razonable RR, Asberg A, Rollag H, et al. Virologic suppression measured by a CMV DNA test calibrated to the WHO international standard is predictive of CMV disease resolution in transplant recipients. Clin Infect Dis 2013 [Epub ahead of print].

Razonable RR, Brown RA, Wilson J, et al. The clinical use of various blood compartments for cytomegalovirus (CMV) DNA quantitation in transplant recipients with CMV disease. Transplantation 2002; 73:968–73.

Razonable RR, Zerr DM; AST Infectious Diseases Community of Practice. HHV-6, HHV-7 and HHV-8 in solid organ transplant recipients. Am J Transplant 2009;(9 Suppl 4):S97–103.

Rupnik M, Wilcox MH, Gerding DN. Clostridium difficile infection: new developments in epidemiology and pathogenesis. Nat Rev Microbiol 2009;7:526–36.

Saslow D, Solomon D, Lawson HW, et al. American Cancer Society, American Society for Colposcopy and Cervical Pathology and American Society for Clinical Pathology screening guidelines for the prevention and early detection of cervical cancer. Am J Clin Pathol 2012;137:516–42.

Schiffman M, Glass AG, Wentzensen N, et al. A long-term prospective study of type-specific human papillomavirus infection and risk of cervical neoplasia among 20,000 women in the Portland Kaiser Cohort Study. Cancer Epidemiol Biomarkers Prev 2011;20:1398–409.

Schwebke JR, Hobbs MM, Taylor SN, et al. Molecular testing for Trichomonas vaginalis in women: results from a prospective U.S. clinical trial. J Clin Microbiol 2011;49:4106–11.

Smith KR, Suppiah V, O'Connor K, et al. Identification of improved IL28B SNPs and haplotypes for prediction of drug response in treatment of hepatitis C using massively paralleled sequencing in a cross-sectional European cohort. Genome Med 2011;3:57.

Stamper PD, Louie L, Wong H, et al. Genotypic and phenotypic characterization of methicillin-susceptible Staphylococcus aureus isolates misidentified as methicillin-resistant Staphylococcus aureus by the BD GeneOhm MRSA assay. J Clin Microbiol 2011;49:1240–4.

Steiner I, Budka H, Chaudhuri A, et al. Viral meningoencephalitis: a review of diagnostic methods and guidelines for management. Eur J Neurol 2010;17:999–1009.

Stoler MH, Wright TC Jr, Sharma A, et al. High-risk human papillomavirus testing in women with ASC-US cytology: results from the ATHENA HPV study. Am J Clin Pathol 2011;135:468–75.

Stoler MH, Wright TC, Cuzick J, et al. Aptima HPV performance in women with atypical squamous cells of undetermined significance cytology results. Am J Obstet Gynecol 2013;144:e1–8.

Thompson MA, Aberg JA, Hoy JF, et al. Antiretroviral treatment of adult HIV infection. 2012 Recomendations of the International Antiviral Society-USA Panel. JAMA 2012;308:387–402.

Valsamakis A. Molecular testing in the diagnosis and management of chronic hepatitis B. Clin Microbiol Rev 2007;20:426–39.

Wolthers KC, Benschop KS, Schinkel J, et al. Human parechoviruses as an important viral cause of sepsis-like illness and meningitis in young children. Clin Infect Dis 2008;47:358–63.

Workowski KA, Berman S. Sexually transmitted diseases treatment guidelines. MMWR Recomm Rep 2010;59:1–110.

World Health Organization. Automated real-time nucleic acid amplification technique for rapid and simultaneous detection of tuberculosis and rifampin resistance [Policy Statement]. 2011.

Wright TC, Massad LS, Dunton CJ, et al. 2006 Consensus guidelines for the management of women with abnormal cervical cancer screening tests. Am J Obstet Gynecol 2007;197:346–55.

Zuckerman RA, Limaye AP. Varicella zoster virus (VZV) and herpes simplex virus (HSV) in solid organ transplant patients. Am J Transplant 2013;(13 Suppl 3):55–66.

Index

A

B

C

G

H

I

J

K

L

M

N

O

P

T

U

V

W

X